Fundamentals of FINITE MATHEMATICS

Fundamentals of FINITE MATHEMATICS

Sarah Cook
Joe Swann

KNOWLEDGE BAKERS

Cataloging in Publication Data--DK
Courtesy: D.K. Agencies (P) Ltd. <docinfo@dkagencies.com>

Cook, Sarah (Writer on mathematics), author.
Fundamentals of finite mathematics / Sarah Cook, Joe Swann.
pages cm
Includes bibliographical references and index.
ISBN 978-93-49964-23-5

1. Mathematics. I. Swann, Joe, author. II. Title.

LCC QA39.2.C66 2023 | DDC 519 23

Published by : **Knowledge Bakers**
Pune, India
Email: knowledgebakers2019@gmail.com

Preface

What is finite mathematics? Though there is no easy answer to this question, a brief answer would include “mathematics that works outside of the notion of continuity.” The realm of continuity is the realm of calculus. Consequently, mathematical modelling of real life problems that are limited to discrete (finite packs of) data or information falls into realm of finite mathematics. Hence, computer scientists should not miss taking finite mathematics since by its very nature a computer works with discrete data. However, finite mathematics goes further than discrete mathematics even though they are closely related. Finite mathematics is more about learning mathematical modelling techniques that can be applied to real world problems. It is related to almost all other areas of mathematics and it also has a wide range of applications. Modern high-speed computers have made all forms of computation vastly in an easy form. In past, while the calculations used to take several days to complete, these can now be carried out in a few seconds, with the help of computers. The availability of modern computers can take a great deal of dudgery, out of mathematical computations and makes possible large-scale computations that would have been impossible, otherwise.

The term finite mathematics is sometimes applied to parts of the field of discrete mathematics that deals with finite sets, particularly those areas relevant to business. For example, a system of linear equations is a modelling technique that can be applied to many real world situations, including business and economics, finance, life sciences, social sciences, computer sciences and the physical sciences. Hence, in finite mathematics we learn methods to solve systems of linear equations. Of course, not all situations in the real world are bound by equality and therefore we need to learn how to model using inequalities and how to solve a system of linear inequalities. These techniques form the basis of linear programming. Techniques learned in finite mathematics are the basis of Operations Research that has applications in transportation, shipping, scheduling and optimization. Basic financial mathematics is also in the realm of finite mathematics with a particular emphasis in compound interest, future values of annuities, present values of annuities and amortization. Probability and statistics have to work with discrete distributions and the basis for these fields is set theory. A function defined on an interval of the integers is usually called a sequence. A sequence could be a finite sequence from a data source or an infinite sequence from a discrete dynamical system. Such a discrete function could be defined explicitly by a list (if its domain is finite), or by a formula for its general term, or it could be given implicitly by a recurrence relation or difference equation. Difference equations are similar to a differential equations, but replace differentiation by taking the difference between adjacent terms; they can be used to approximate differential equations or (more often) studied in their own right. Many questions and methods concerning differential equations have counterparts for difference equations.

Finite mathematics represents an important trend in mathematics. It is possible to single out its typical subject of study, methods, and problems, whose nature is largely determined by the necessity, characteristic of finite mathematics, for rejecting the fundamental concepts of classical mathematics—limit and continuity—and by the fact that the powerful methods of classical mathematics as a rule prove to be of little use in many problems of finite mathematics. In addition to delimiting finite mathematics by indicating its subject of study, it is also possible to define it by specifying its subdivisions. These include combinatorial analysis, the theory of graphs, coding theory, and the theory of functional systems. Viewed in these terms, finite mathematics stands for the study of finite structures. Viewed in less restrictive terms, finite mathematics includes both entire branches of mathematics, such as mathematical logic, and parts of these branches, such as number theory, algebra, computer science, and discrete probability theory.

The elements of finite mathematics arose in early antiquity. They developed parallel to other branches of mathematics and were to a significant extent a part of them. Typical for that period were problems that were related to the properties of integers and subsequently led to the creation of number theory.

The present book deals with all the important dimensions of this subject. It is a valuable reference source for all those concerned with this subject.

Sarah Cook
Joe Swann

CONTENTS

Chapter

1

Variants and Generalizations

Alternative Definition of a Function

The above definition of "a function from X to Y" is generally agreed on, however there are two different ways a "function" is normally defined where the domain X and codomain Y are not explicitly or implicitly specified. Usually this is not a problem as the domain and codomain normally will be known. With one definition saying the function defined by $f(x) = x^2$ on the reals does not completely specify a function as the codomain is not specified, and in the other it is a valid definition.

In the other definition a function is defined as a set of ordered pairs where each first element only occurs once. The domain is the set of all the first elements of a pair and there is no explicit codomain separate from the image. Concepts like surjective have to be refined for such functions, more specifically by saying that a (given) function is *surjective on a (given) set* if its image equals that set.

If a function is defined as a set of ordered pairs with no specific codomain, then $f: X \rightarrow Y$ indicates that f is a function whose domain is X and whose image is a subset of Y.Y may be referred to as the codomain but then any set including the image of f is a valid codomain of f. This is also referred to by saying that "f maps X into Y" In some usages X and Y may subset the ordered pairs, e.g. the function f on the real numbers such that $y=x^2$ when used as in f: [0,4] → [0,4] means the function defined only on the interval [0,2]. With the definition of a function as an ordered triple this would always be considered a partial function.

An alternative definition of the composite function $g(f(x))$ defines it for the set of all x in the domain of f such that $f(x)$ is in the domain of g. Thus the real square root of $-x^2$ is a function only defined at 0 where it has the value 0. When the definition of a function by its graph only is used, since the codomain is not defined, the "surjection" must be accompanied with a statement about the set the function maps onto. For example, we might say f maps onto the set of all real numbers.

Functions are commonly defined as a type of relation. A relation from X to Y is a set of ordered pairs(x, y) with $x \in X$ and $y \in Y$. A function from X to Y can be described as a relation from X to Y that is left-total and right-unique. However when X and Y are not specified there is a disagreement about the definition of a relation that parallels that for functions. Normally a relation is just defined as a set of ordered pairs and a correspondence is defined as a triple (X, Y, F), however the distinction between the two is often blurred or a relation is never referred to without specifying the two sets. The definition of a function as a triple defines a function as a type of correspondence, whereas the definition of a function as an ordered pair defines a function as a type of relation.

Many operations in set theory, such as the power set, have the class of all sets as their domain, and therefore, although they are informally described as functions, they do not fit the set-theoretical definition outlined above, because a class is not necessarily a set. However some definitions of relations and functions define them as classes of pairs rather than sets of pairs and therefore do include the power set as a function.

Partial and Multi-valued Functions

$f(x) = \pm\sqrt{x}$ is not a function in the proper sense, but a multi-valued function: it assigns to each positive real number x two values: the (positive) square root of x, and $-\sqrt{x}$.

In some parts of mathematics, including recursion theory and functional analysis, it is convenient to study *partial functions* in which some values of the domain have no association in the graph; i.e., single-valued relations. For example, the function f such that $f(x) = 1/x$ does not define a value for $x = 0$, since division by zero is not defined. Hence f is only a partial function from the real line to the real line. The term total function can be used to stress the fact that every element of the domain does appear as the first element of an ordered pair in the graph. In other parts of mathematics, non-single-valued relations are similarly conflated with functions: these are called *multivalued functions*, with the corresponding term single-valued function for ordinary functions.

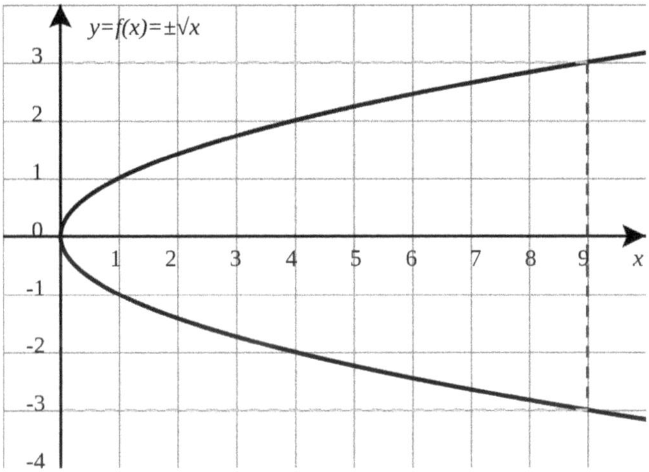

Functions with Multiple Inputs and Outputs

The concept of function can be extended to an object that takes a combination of two (or more) argument values to a single result. This intuitive concept is formalized by a function whose domain is the Cartesian product of two or more sets.

For example, consider the function that associates two integers to their product: $f(x, y) = x \cdot y$. This function can be defined formally as having domain Z×Z, the set of all integer pairs; codomain Z; and, for graph, the set of all pairs $((x,y), x \cdot y)$. Note that the first component of any such pair is itself a pair (of integers), while the second component is a single integer.

The function value of the pair (x,y) is $f((x,y))$. However, it is customary to drop one set of parentheses and consider $f(x,y)$ a function of two variables, x and y. Functions of two variables may be plotted on the three-dimensional Cartesian as ordered triples of the form $(x,y,f(x,y))$.

The concept can still further be extended by considering a function that also produces output that is expressed as several variables. For example, consider the integer divide function, with domain $Z \times N$ and codomain $Z \times N$. The resultant (quotient, remainder) pair is a single value in the codomain seen as a Cartesian product.

Binary Operations

The familiar binary operations of arithmetic, addition and multiplication, can be viewed as functions from $R \times R$ to R. This view is generalized in abstract algebra, where n-ary functions are used to model the operations of arbitrary algebraic structures. For example, an abstract group is defined as a set X and a function f from $X \times X$ to X that satisfies certain properties.

Traditionally, addition and multiplication are written in the infix notation: $x + y$ and $x \times y$ instead of $+(x, y)$ and $\times(x, y)$.

Functors

The idea of structure-preserving functions, or homomorphisms, led to the abstract notion of morphism, the key concept of category theory. In fact, functions $f: X \to Y$ are the morphisms in the category of sets, including the empty set: if the domain X is the empty set, then the subset of $X \times Y$ describing the function is necessarily empty, too. However, this is still a well-defined function. Such a function is called an empty function. In particular, the identity function of the empty set is defined, a requirement for sets to form a category.

The concept of categorification is an attempt to replace set-theoretic notions by category-theoretic ones. In particular, according to this idea, sets are replaced by categories, while functions between sets are replaced by functors.

Linear Regression

In statistics, linear regression is an approach to modelling the relationship between a scalar dependent variable$y$ and one or more explanatory variables denoted X. The case of one explanatory variable is called *simple linear regression*. For more than one explanatory variable, it is called *multiple linear regression*. (This term should be distinguished from *multivariate linear regression*, where multiple correlated dependent variables are predicted, rather than a single scalar variable.)

In linear regression, data are modelled using linear predictor functions, and unknown model parameters are estimated from the data. Such models are called *linear models*. Most commonly, linear regression refers to a model in which the conditional mean of y given the value of X is an affine function of X. Less commonly, linear regression could refer to a model in which the median, or some other quantile of the conditional distribution of y given X is expressed as a linear function of X. Like all forms of regression analysis, *linear regression* focuses on the conditional probability distribution of y given X, rather than on the joint probability distribution of y and X, which is the domain of multivariate analysis.

Linear regression was the first type of regression analysis to be studied rigorously, and to be used extensively in practical applications. This is because models which depend linearly on their unknown parameters are easier to fit than models which are non-linearly related to their parameters and because the statistical properties of the resulting estimators are easier to determine.

Linear regression has many practical uses. Most applications fall into one of the following two broad categories:

- If the goal is prediction, or forecasting, linear regression can be used to fit a predictive model to an observed data set of y and X values. After developing such a model, if an additional value of X is then given without its accompanying value of y, the fitted model can be used to make a prediction of the value of y.
- Given a variable y and a number of variables $X_1, \ldots, X_p$ that may be related to y, linear regression analysis can be applied to quantify the strength of the relationship between y and the X_j, to assess which X_j may have no relationship with y at all, and to identify which subsets of the X_j contain redundant information about y.

Linear regression models are often fitted using the least squares approach, but they may also be fitted in other ways, such as by minimizing the "lack of fit" in some other norm (as with least absolute deviations regression), or by minimizing a penalized version of the least squares loss function as in ridge regression. Conversely, the least squares approach can be used to fit models that are not linear models. Thus, although the terms "least squares" and "linear model" are closely linked, they are not synonymous.

Introduction to Linear Regression

Given a data set $\{y_i, x_{i1}, \ldots, x_{ip}\}_{i=1}^{n}$ of n statistical units, a linear regression model assumes that the relationship between the dependent variable y_i and the p-vector of regressors x_i is linear. This relationship is modelled through a *disturbance term* or *error variable* ε_i — an unobserved random variable that adds noise to the linear relationship between the dependent variable and regressors.

Thus the model takes the form

$$y_i = \beta_1 x_{i1} + \cdots + \beta_p x_{ip} + \varepsilon_i = x_i^T \beta + \varepsilon_i, \qquad i = 1, \ldots, n,$$

where ^{T} denotes the transpose, so that $x_i^T\beta$ is the inner product between vectors x_i and β. Often these n equations are stacked together and written in vector form as

$$y = X\beta + \varepsilon,$$

where

$$y = \begin{pmatrix} y_1 \\ y_2 \\ \vdots \\ y_n \end{pmatrix}, \quad X = \begin{pmatrix} x_1^T \\ x_2^T \\ \vdots \\ x_n^T \end{pmatrix} = \begin{pmatrix} x_{11} & \cdots & x_{1p} \\ x_{21} & \cdots & x_{2p} \\ \vdots & \ddots & \vdots \\ x_{n1} & \cdots & x_{np} \end{pmatrix}, \quad \beta = \begin{pmatrix} \beta_1 \\ \vdots \\ \beta_p \end{pmatrix}, \quad \varepsilon = \begin{pmatrix} \varepsilon_1 \\ \varepsilon_2 \\ \vdots \\ \varepsilon_n \end{pmatrix}.$$

Some remarks on terminology and general use:

- y_i is called the *regress and*, *endogenous variable*, *response variable*, *measured variable*, or *dependent variable*. The decision as to which variable in a data set is modelled as the dependent variable and which are modelled as the independent variables may be based on a presumption that the value of one of the variables is caused by, or directly influenced by the other variables. Alternatively, there may be an operational reason to model one of the variables in terms of the others, in which case there need be no presumption of causality.
- x_i are called *regressors*, *exogenous variables*, *explanatory variables*, *covariates*, *input variables*, *predictor variables*, or *independent variables*. The matrix X is sometimes called the design matrix.
  - Usually a constant is included as one of the regressors. For example we can take $x_{i1} = 1$ for $i = 1, \ldots, n$. The corresponding element of β is called the *intercept*. Many statistical inference procedures for linear models require an intercept to be present, so it is often included even if theoretical considerations suggest that its value should be zero.
  - Sometimes one of the regressors can be a non-linear function of another regressor or of the data, as in polynomial regression and segmented regression. The model remains linear as long as it is linear in the parameter vector β.
  - * The regressors x_{ij} may be viewed either as random variables, which we simply observe, or they can be considered as

predetermined fixed values which we can choose. Both interpretations may be appropriate in different cases, and they generally lead to the same estimation procedures; however different approaches to asymptotic analysis are used in these two situations.

- ß is a p-dimensional *parameter vector*. Its elements are also called *effects*, or *regression coefficients*. Statistical estimation and inference in linear regression focuses on β.
- ε_i is called the *error term*, *disturbance term*, or *noise*. This variable captures all other factors which influence the dependent variable y_i other than the regressors x_i. The relationship between the error term and the regressors, for example whether they are correlated, is a crucial step in formulating a linear regression model, as it will determine the method to use for estimation.

Example. Consider a situation where a small ball is being tossed up in the air and then we measure its heights of ascent h_i at various moments in time t_i. Physics tells us that, ignoring the drag, the relationship can be modelled as

$$h_i = \beta_1 t_i + \beta_2 t_i^2 + \varepsilon_i,$$

where β_1 determines the initial velocity of the ball, β_2 is proportional to the standard gravity, and ε_i is due to measurement errors. Linear regression can be used to estimate the values of β_1 and β_2 from the measured data. This model is non-linear in the time variable, but it is linear in the parameters β_1 and β_2; if we take regressors $x_i = (x_{i1}, x_{i2}) = (t_i, t_i^2)$, the model takes on the standard form

$$h_i = x_i^T \beta + \varepsilon_i.$$

Assumptions

Standard linear regression models with standard estimation techniques make a number of assumptions about the predictor variables, the response variables and their relationship. Numerous extensions have been developed that allow each of these assumptions to be relaxed (i.e. reduced to a weaker form), and in some cases eliminated entirely. Some methods are general enough that they can relax multiple assumptions at once, and in other cases this can be achieved by combining different extensions. Generally these extensions make the estimation procedure more complex and time-consuming, and may also require more data in order to get an accurate model.

The following are the major assumptions made by standard linear regression models with standard estimation techniques (e.g. ordinary least squares):

- Weak exogeneity. This essentially means that the predictor variables x can be treated as fixed values, rather than random variables. This means, for example, that the predictor variables are assumed to be error-free, that is they are not contaminated with measurement errors. Although not realistic in many settings, dropping this assumption leads to significantly more difficult errors-in-variables models.
- Linearity. This means that the mean of the response variable is a linear combination of the parameters (regression coefficients) and the predictor variables. Note that this assumption is much less restrictive than it may at first seem. Because the predictor variables are treated as fixed values, linearity is really only a restriction on the parameters. The predictor variables themselves can be arbitrarily transformed, and in fact multiple copies of the same underlying predictor variable can be added, each one transformed differently. This trick is used, for example, in polynomial regression, which uses linear regression to fit the response variable as an arbitrary polynomial function (up to a given rank) of a predictor variable. This makes linear regression an extremely powerful inference method. In fact, models such as polynomial regression are often "too powerful", in that they tend to overfit the data. As a result, some kind of regularization must typically be used to prevent unreasonable solutions coming out of the estimation process. Common examples are ridge regression and lasso regression. Bayesian linear regression can also be used, which by its nature is more or less immune to the problem of overfitting. (In fact, ridge regression and lasso regression can both be viewed as special cases of Bayesian linear regression, with particular types of prior distributions placed on the regression coefficients.)
- Constant variance (aka homoscedasticity). This means that different response variables have the same variance in their errors, regardless of the values of the predictor variables. In practice this assumption is invalid (i.e. the errors are heteroscedastic) if the response variables can vary over a wide scale. In order to determine for heterogeneous error variance, or when a pattern of residuals violates model assumptions of

homoscedasticity (error is equally variable around the 'best-fitting line' for all points of x), it is prudent to look for a "fanning effect" between residual error and predicted values. This is to say there will be a systematic change in the absolute or squared residuals when plotted against the predicting outcome. Error will not be evenly distributed across the regression line. Heteroscedasticity will result in the averaging over of distinguishable variances around the points to get a single variance that is inaccurately representing all the variances of the line. In effect, residuals appear clustered and spread apart on their predicted plots for larger and smaller values for points along the linear regression line, and the mean squared error for the model will be wrong. Typically, for example, a response variable whose mean is large will have a greater variance than one whose mean is small.

For example, a given person whose income is predicted to be $100,000 may easily have an actual income of $80,000 or $120,000 (a standard deviation of around $20,000), while another person with a predicted income of $10,000 is unlikely to have the same $20,000 standard deviation, which would imply their actual income would vary anywhere between -$10,000 and $30,000. (In fact, as this shows, in many cases – often the same cases where the assumption of normally distributed errors fails – the variance or standard deviation should be predicted to be proportional to the mean, rather than constant.) Simple linear regression estimation methods give less precise parameter estimates and misleading inferential quantities such as standard errors when substantial heteroscedasticity is present. However, various estimation techniques (e.g. weighted least squares and heteroscedasticity-consistent standard errors) can handle heteroscedasticity in a quite general way. Bayesian linear regression techniques can also be used when the variance is assumed to be a function of the mean. It is also possible in some cases to fix the problem by applying a transformation to the response variable (e.g. fit the logarithm of the response variable using a linear regression model, which implies that the response variable has a log-normal distribution rather than a normal distribution).

- Independence of errors. This assumes that the errors of the response variables are uncorrelated with each other. (Actual statistical independence is a stronger condition than mere lack of correlation and is often not needed, although it can be exploited if it is known to hold.) Some methods (e.g. generalized least squares) are capable of handling correlated errors, although they typically require significantly more data unless some sort of regularization is used to bias the model towards assuming uncorrelated errors. Bayesian linear regression is a general way of handling this issue.

- Lack of multicollinearity in the predictors. For standard least squares estimation methods, the design matrix X must have full column rank p,; otherwise, we have a condition known as multicollinearity in the predictor variables. This can be triggered by having two or more perfectly correlated predictor variables (e.g. if the same predictor variable is mistakenly given twice, either without transforming one of the copies or by transforming one of the copies linearly). It can also happen if there is too little data available compared to the number of parameters to be estimated (e.g. fewer data points than regression coefficients). In the case of multicollinearity, the parameter vector β will be non-identifiable — it has no unique solution. At most we will be able to identify some of the parameters, i.e. narrow down its value to some linear subspace of R^p. Methods for fitting linear models with multicollinearity have been developed; some require additional assumptions such as "effect sparsity" — that a large fraction of the effects are exactly zero. Note that the more computationally expensive iterated algorithms for parameter estimation, such as those used in generalized linear models, do not suffer from this problem — and in fact it's quite normal to when handling categorically-valued predictors to introduce a separate indicator variable predictor for each possible category, which inevitably introduces multicollinearity.

Beyond these assumptions, several other statistical properties of the data strongly influence the performance of different estimation methods:

- The statistical relationship between the error terms and the regressors plays an important role in determining whether an estimation procedure has desirable sampling properties such as being unbiased and consistent.
- The arrangement, or probability distribution of the predictor variables x has a major influence on the precision of estimates of β. Sampling and design of experiments are highly-developed

subfields of statistics that provide guidance for collecting data in such a way to achieve a precise estimate of β.

Interpretation

A fitted linear regression model can be used to identify the relationship between a single predictor variable x_j and the response variable y when all the other predictor variables in the model are "held fixed". Specifically, the interpretation of β_j is the expected change in y for a one-unit change in x_j when the other covariates are held fixed—that is, the expected value of the partial derivative of y with respect to x_j. This is sometimes called the *unique effect* of x_j on y. In contrast, the *marginal effect* of x_j on y can be assessed using a correlation coefficient or simple linear regression model relating x_j to y; this effect is the total derivative of y with respect to x_j.

Care must be taken when interpreting regression results, as some of the regressors may not allow for marginal changes (such as dummy variables, or the intercept term), while others cannot be held fixed (recall the example from the introduction: it would be impossible to "hold t_i fixed" and at the same time change the value of t_i^2).

It is possible that the unique effect can be nearly zero even when the marginal effect is large. This may imply that some other covariate captures all the information in x_j, so that once that variable is in the model, there is no contribution of x_j to the variation in y. Conversely, the unique effect of x_j can be large while its marginal effect is nearly zero. This would happen if the other covariates explained a great deal of the variation of y, but they mainly explain variation in a way that is complementary to what is captured by x_j. In this case, including the other variables in the model reduces the part of the variability of y that is unrelated to x_j, thereby strengthening the apparent relationship with x_j.

The meaning of the expression "held fixed" may depend on how the values of the predictor variables arise. If the experimenter directly sets the values of the predictor variables according to a study design, the comparisons of interest may literally correspond to comparisons among units whose predictor variables have been "held fixed" by the experimenter. Alternatively, the expression "held fixed" can refer to a selection that takes place in the context of data analysis. In this case, we "hold a variable fixed" by restricting our attention to the subsets of the data that happen to have a common value for the given predictor variable. This is the only interpretation of "held fixed" that can be used in an observational study.

The notion of a "unique effect" is appealing when studying a complex system where multiple interrelated components influence the response variable.

In some cases, it can literally be interpreted as the causal effect of an intervention that is linked to the value of a predictor variable. However, it has been argued that in many cases multiple regression analysis fails to clarify the relationships between the predictor variables and the response variable when the predictors are correlated with each other and are not assigned following a study design.

Extensions

Numerous extensions of linear regression have been developed, which allow some or all of the assumptions underlying the basic model to be relaxed.

Simple and Multiple Regression

The very simplest case of a single scalar predictor variable x and a single scalar response variable y is known as *simple linear regression*. The extension to multiple and/or vector-valued predictor variables (denoted with a capital X) is known as *multiple linear regression*. Nearly all real-world regression models involve multiple predictors, and basic descriptions of linear regression are often phrased in terms of the multiple regression model.

Note, however, that in these cases the response variable y is still a scalar.

General Linear Models

The general linear model considers the situation when the response variable Y is not a scalar but a vector. Conditional linearity of $E(y \mid x) = Bx$ is still assumed, with a matrix B replacing the vector β of the classical linear regression model. Multivariate analogues of OLS and GLS have been developed.

Heteroscedastic Models

Various models have been created that allow for heteroscedasticity, i.e. the errors for different response variables may have different variances. For example, weighted least squares is a method for estimating linear regression models when the response variables may have different error variances, possibly with correlated errors. Heteroscedasticity-consistent standard errors is an improved method for use with uncorrelated but potentially heteroscedastic errors.

Generalized Linear Models

Generalized linear models (GLM's) are a framework for modelling a response variable *y* that is bounded or discrete. This is used, for example:

- when modelling positive quantities (e.g. prices or populations) that vary over a large scale — which are better described using a skewed distribution such as the log-normal distribution or Poisson distribution (although GLM's are not used for log-normal data, instead the response variable is simply transformed using the logarithm function);
- when modelling categorical data, such as the choice of a given candidate in an election (which is better described using a Bernoulli distribution/binomial distribution for binary choices, or a categorical distribution/multinomial distribution for multi-way choices), where there are a fixed number of choices that cannot be meaningfully ordered;
- when modelling ordinal data, e.g. ratings on a scale from 0 to 5, where the different outcomes can be ordered but where the quantity itself may not have any absolute meaning (e.g. a rating of 4 may not be "twice as good" in any objective sense as a rating of 2, but simply indicates that it is better than 2 or 3 but not as good as 5).

Generalized linear models allow for an arbitrary *link function g* that relates the mean of the response variable to the predictors, i.e. $E(y) = g(\beta 2\ x)$. The link function is often related to the distribution of the response, and in particular it typically has the effect of transforming between the $(-\infty,\infty)$ range of the linear predictor and the range of the response variable.

Some common examples of GLM's are:

- Poisson regression for count data.
- Logistic regression and probit regression for binary data.
- Multinomial logistic regression and multinomial probit regression for categorical data.
- Ordered probit regression for ordinal data.

Single index models allow some degree of nonlinearity in the relationship between *x* and *y*, while preserving the central role of the linear predictor $\beta 2\ x$ as in the classical linear regression model. Under certain conditions, simply applying OLS to data from a single-index model will consistently estimate β up to a proportionality constant.

Hierarchical Linear Models

Hierarchical linear models (or *multilevel regression*) organizes the data into a hierarchy of regressions, for example where A is regressed on B, and B is regressed on C. It is often used where the data have a natural hierarchical structure such as in educational statistics, where students are nested in classrooms, classrooms are nested in schools, and schools are nested in some administrative grouping such as a school district.

The response variable might be a measure of student achievement such as a test score, and different covariates would be collected at the classroom, school, and school district levels.

Errors-in-variables

Errors-in-variables models (or "measurement error models") extend the traditional linear regression model to allow the predictor variables X to be observed with error. This error causes standard estimators of β to become biased. Generally, the form of bias is an attenuation, meaning that the effects are biased towards zero.

Others

- In Dempster–Shafer theory, or a linear belief function in particular, a linear regression model may be represented as a partially swept matrix, which can be combined with similar matrices representing observations and other assumed normal distributions and state equations. The combination of swept or unswept matrices provides an alternative method for estimating linear regression models.

Estimation Methods

A large number of procedures have been developed for parameter estimation and inference in linear regression. These methods differ in computational simplicity of algorithms, presence of a closed-form solution, robustness with respect to heavy-tailed distributions, and theoretical assumptions needed to validate desirable statistical properties such as consistency and asymptotic efficiency.

Some of the more common estimation techniques for linear regression are summarized below.

Least-squares Estimation and Related Techniques

- Ordinary least squares (OLS) is the simplest and thus most common estimator. It is conceptually simple and computationally straightforward. OLS estimates are commonly used to analyze both experimental and observational data.

The OLS method minimizes the sum of squared residuals, and leads to a closed-form expression for the estimated value of the unknown parameter β:

$$\hat{\beta} = (X^T X)^{-1} X^T y = \left(\tfrac{1}{n}\sum x_i x_i^T\right)^{-1}\left(\tfrac{1}{n}\sum x_i y_i\right).$$

The estimator is unbiased and consistent if the errors have finite variance and are uncorrelated with the regressors

$$E[x_i \varepsilon_i] = 0.$$

It is also efficient under the assumption that the errors have finite variance and are homoscedastic, meaning that $E[\varepsilon_i^2 \mid x_i]$ does not depend on i. The condition that the errors are uncorrelated with the regressors will generally be satisfied in an experiment, but in the case of observational data, it is difficult to exclude the possibility of an omitted covariate z that is related to both the observed covariates and the response variable. The existence of such a covariate will generally lead to a correlation between the regressors and the response variable, and hence to an inconsistent estimator of β. The condition of homoscedasticity can fail with either experimental or observational data. If the goal is either inference or predictive modelling, the performance of OLS estimates can be poor if multicollinearity is present, unless the sample size is large.

In simple linear regression, where there is only one regressor (with a constant), the OLS coefficient estimates have a simple form that is closely related to the correlation coefficient between the covariate and the response.

- Generalized least squares (GLS) is an extension of the OLS method, that allows efficient estimation of β when either heteroscedasticity, or correlations, or both are present among the error terms of the model, as long as the form of heteroscedasticity and correlation is known independently of the data. To handle heteroscedasticity when the error terms are uncorrelated with each other, GLS minimizes a weighted analogue to the sum of squared residuals from OLS regression, where the weight for the ith case is inversely proportional to $\mathrm{var}(\varepsilon_i)$. This special case of GLS is called "weighted least squares". The GLS solution to estimation problem is

$$\hat{\beta} = (X^T \Omega^{-1} X)^{-1} X^T \Omega^{-1} y,$$

where Ω is the covariance matrix of the errors. GLS can be viewed as applying a linear transformation to the data so that the assumptions of OLS are met for the transformed data. For GLS to be applied, the covariance structure of the errors must be known up to a multiplicative constant.

- Percentage least squares focuses on reducing percentage errors, which is useful in the field of forecasting or time series analysis. It is also useful in situations where the dependent variable has a wide range without constant variance, as here the larger residuals at the upper end of the range would dominate if OLS were used. When the percentage or relative error is normally distributed, least squares percentage regression provides maximum likelihood estimates. Percentage regression is linked to a multiplicative error model, whereas OLS is linked to models containing an additive error term.
- Iteratively reweighted least squares (IRLS) is used when heteroscedasticity, or correlations, or both are present among the error terms of the model, but where little is known about the covariance structure of the errors independently of the data. In the first iteration, OLS, or GLS with a provisional covariance structure is carried out, and the residuals are obtained from the fit. Based on the residuals, an improved estimate of the covariance structure of the errors can usually be obtained. A subsequent GLS iteration is then performed using this estimate of the error structure to define the weights. The process can be iterated to convergence, but in many cases, only one iteration is sufficient to achieve an efficient estimate of β.
- Instrumental variables regression (IV) can be performed when the regressors are correlated with the errors. In this case, we need the existence of some auxiliary *instrumental variables*z_i such that $E[z_i\varepsilon_i] = 0$. If Z is the matrix of instruments, then the estimator can be given in closed form as

$$\hat{\beta} = (X^T Z(Z^T Z)^{-1} Z^T X)^{-1} X^T Z(Z^T Z)^{-1} Z^T y.$$

- Optimal instruments regression is an extension of classical IV regression to the situation where $E[\varepsilon_i \mid z_i] = 0$.
- Total least squares (TLS) is an approach to least squares estimation of the linear regression model that treats the covariates and response variable in a more geometrically symmetric manner than OLS. It is one approach to handling

the "errors in variables" problem, and is sometimes used when the covariates are assumed to be error-free.

Maximum-likelihood Estimation and Related Techniques

- Maximum likelihood estimation can be performed when the distribution of the error terms is known to belong to a certain parametric family f_θ of probability distributions. When f_θ is a normal distribution with mean zero and variance θ, the resulting estimate is identical to the OLS estimate. GLS estimates are maximum likelihood estimates when ε follows a multivariate normal distribution with a known covariance matrix.
- Ridge regression, and other forms of penalized estimation such as Lasso regression, deliberately introduce bias into the estimation of β in order to reduce the variability of the estimate. The resulting estimators generally have lower mean squared error than the OLS estimates, particularly when multicollinearity is present. They are generally used when the goal is to predict the value of the response variable y for values of the predictors x that have not yet been observed. These methods are not as commonly used when the goal is inference, since it is difficult to account for the bias.
- Least absolute deviation (LAD) regression is a robust estimation technique in that it is less sensitive to the presence of outliers than OLS (but is less efficient than OLS when no outliers are present). It is equivalent to maximum likelihood estimation under a Laplace distribution model for ε.

Adaptive estimation. If we assume that error terms are independent from the regressors $\varepsilon_i \perp x_i$, the optimal estimator is the 2-step MLE, where the first step is used to non-parametrically estimate the distribution of the error term.

Other Estimation Techniques

- Bayesian linear regression applies the framework of Bayesian statistics to linear regression. In particular, the regression coefficients β are assumed to be random variables with a specified prior distribution. The prior distribution can bias the solutions for the regression coefficients, in a way similar to (but more general than) ridge regression or lasso regression. In addition, the Bayesian estimation process produces not a single point estimate for the "best" values of the regression coefficients but an entire posterior distribution, completely describing the uncertainty surrounding the quantity. This can be used to estimate the "best" coefficients using the mean, mode, median, any quantile, or any other function of the posterior distribution.

- Quantile regression focuses on the conditional quantiles of y given X rather than the conditional mean of y given X. Linear quantile regression models a particular conditional quantile, often the conditional median, as a linear function $\beta^T x$ of the predictors.
- Mixed models are widely used to analyze linear regression relationships involving dependent data when the dependencies have a known structure. Common applications of mixed models include analysis of data involving repeated measurements, such as longitudinal data, or data obtained from cluster sampling. They are generally fit as parametric models, using maximum likelihood or Bayesian estimation. In the case where the errors are modelled as normal random variables, there is a close connection between mixed models and generalized least squares. Fixed effects estimation is an alternative approach to analyzing this type of data.
- Principal component regression (PCR) is used when the number of predictor variables is large, or when strong correlations exist among the predictor variables. This two-stage procedure first reduces the predictor variables using principal component analysis then uses the reduced variables in an OLS regression fit. While it often works well in practice, there is no general theoretical reason that the most informative linear function of the predictor variables should lie among the dominant principal components of the multivariate distribution of the predictor variables. The partial least squares regression is the extension of the PCR method which does not suffer from the mentioned deficiency.
- Least-angle regression is an estimation procedure for linear regression models that was developed to handle high-dimensional covariate vectors, potentially with more covariates than observations.
- The Theil–Sen estimator is a simple robust estimation technique that chooses the slope of the fit line to be the median of the slopes of the lines through pairs of sample points. It has similar statistical efficiency properties to simple linear regression but is much less sensitive to outliers.

- Other robust estimation techniques, including the α-trimmed mean approach, and L-, M-, S-, and R-estimators have been introduced.

Further Discussion

In statistics, the problem of numerical methods for linear least squares is an important one because linear regression models are one of the most important types of model, both as formal statistical models and for exploration of data sets. The majority of statistical computer packages contain facilities for regression analysis that make use of linear least squares computations. Hence it is appropriate that considerable effort has been devoted to the task of ensuring that these computations are undertaken efficiently and with due regard to numerical precision.

Individual statistical analyses are seldom undertaken in isolation, but rather are part of a sequence of investigatory steps. Some of the topics involved in considering numerical methods for linear least squares relate to this point. Thus important topics can be

- Computations where a number of similar, and often nested, models are considered for the same data set. That is, where models with the same dependent variable but different sets of independent variables are to be considered, for essentially the same set of data points.
- Computations for analyses that occur in a sequence, as the number of data points increases.
- Special considerations for very extensive data sets.

Fitting of linear models by least squares often, but not always, arises in the context of statistical analysis. It can therefore be important that considerations of computational efficiency for such problems extend to all of the auxiliary quantities required for such analyses, and are not restricted to the formal solution of the linear least squares problem. Matrix calculations, like any others, are affected by rounding errors. An early summary of these effects, regarding the choice of computational methods for matrix inversion, was provided by Wilkinson.

Applications of Linear Regression

Linear regression is widely used in biological, behavioural and social sciences to describe possible relationships between variables. It ranks as one of the most important tools used in these disciplines.

Trend Line

A trend line represents a trend, the long-term movement in time series data after other components have been accounted for. It tells whether a particular data set (say GDP, oil prices or stock prices) have increased or decreased over the period of time. A trend line could simply be drawn by eye through a set of data points, but more properly their position and slope is calculated using statistical techniques like linear regression. Trend lines typically are straight lines, although some variations use higher degree polynomials depending on the degree of curvature desired in the line.

Trend lines are sometimes used in business analytics to show changes in data over time. This has the advantage of being simple. Trend lines are often used to argue that a particular action or event (such as training, or an advertising campaign) caused observed changes at a point in time. This is a simple technique, and does not require a control group, experimental design, or a sophisticated analysis technique. However, it suffers from a lack of scientific validity in cases where other potential changes can affect the data.

Epidemiology

Early evidence relating tobacco smoking to mortality and morbidity came from observational studies employing regression analysis. In order to reduce spurious correlations when analyzing observational data, researchers usually include several variables in their regression models in addition to the variable of primary interest. For example, suppose we have a regression model in which cigarette smoking is the independent variable of interest, and the dependent variable is lifespan measured in years. Researchers might include socio-economic status as an additional independent variable, to ensure that any observed effect of smoking on lifespan is not due to some effect of education or income. However, it is never possible to include all possible confounding variables in an empirical analysis.

For example, a hypothetical gene might increase mortality and also cause people to smoke more. For this reason, randomized controlled trials are often able to generate more compelling evidence of causal relationships than can be obtained using regression analyses of observational data. When controlled experiments are not feasible, variants of regression analysis such as instrumental variables regression may be used to attempt to estimate causal relationships from observational data.

Finance

The capital asset pricing model uses linear regression as well as the concept of Beta for analyzing and quantifying the systematic risk of an investment.

This comes directly from the Beta coefficient of the linear regression model that relates the return on the investment to the return on all risky assets.

Economics

Linear regression is the predominant empirical tool in economics. For example, it is used to predict consumption spending, fixed investment spending, inventory investment, purchases of a country's exports, spending on imports, the demand to hold liquid assets, labour demand, and labour supply.

Environmental Science

Linear regression finds application in a wide range of environmental science applications. In Canada, the Environmental Effects Monitoring Programme uses statistical analyses on fish and benthic surveys to measure the effects of pulp mill or metal mine effluent on the aquatic ecosystem.

Linear Function

In mathematics, the term linear function refers to a function that satisfies the following two properties:

$$f(x+y)=f(x)+f(y)$$
$$f(ax)=af(x).$$

The linear functions may be confused with affine functions. One variable affine functions can be written as $f(x)=mx+b$. Although affine functions make lines when graphed, they do not satisfy the properties of linearity.

Vector Spaces

In mathematics, a *linear function* means a function that is a linear map, that is, a map between two vector spaces that preserves vector addition and scalar multiplication. For example, if x and $f(x)$ are represented as coordinate vectors, then the linear functions are those functions f that can be expressed as

$$f(x)=\mathrm{M}x,$$

where M is a matrix. A function

$$f(x) = mx + b$$

is a linear map if and only if b = 0. For other values of this falls in the more general class of affine maps.

Linear functions form the basis of linear algebra.

Linear Model

In statistics, the term linear model is used in different ways according to the context. The most common occurrence is in connection with regression models and the term is often taken as synonymous with linear regression model. However, the term is also used in time series analysis with a different meaning. In each case, the designation "linear" is used to identify a subclass of models for which substantial reduction in the complexity of the related statistical theory is possible.

Linear Regression Models

For the regression case, the statistical model is as follows. Given a (random) sample $(Y_i, X_{i1}, \ldots, X_{ip}), i = 1, \ldots, n$ the relation between the observations Y_i and the independent variables X_{ij} is formulated as

$$Y_i = \beta_0 + \beta_1 \phi_1(X_{i1}) + \cdots + \beta_p \phi_p(X_{ip}) + \varepsilon_i \qquad i = 1, \ldots, n$$

where $\phi_1, \ldots, \phi_p$ may be nonlinear functions. In the above, the quantities ε_i are random variables representing errors in the relationship. The "linear" part of the designation relates to the appearance of the regression coefficients, β_j in a linear way in the above relationship. Alternatively, one may say that the predicted values corresponding to the above model, namely

$$\hat{Y}_i = \beta_0 + \beta_1 \phi_1(X_{i1}) + \cdots + \beta_p \phi_p(X_{ip}) \qquad (i = 1, \ldots, n),$$

are linear functions of the β_j.

Given that estimation is undertaken on the basis of a least squares analysis, estimates of the unknown parameters β_j are determined by minimising a sum of squares function

$$S = \sum_{i=1}^{n} \left(Y_i - \beta_0 - \beta_1 \phi_1(X_{i1}) - \cdots - \beta_p \phi_p(X_{ip})\right)^2.$$

From this, it can readily be seen that the "linear" aspect of the model means the following:

- the function to be minimised is a quadratic function of the β_j for which minimisation is a relatively simple problem;

- the derivatives of the function are linear functions of the β_j making it easy to find the minimising values;
- the minimising values β_j are linear functions of the observations Y_i;
- the minimising values β_j are linear functions of the random errors ε_i which makes it relatively easy to determine the statistical properties of the estimated values of β_j.

Time Series Models

An example of a linear time series model is an autoregressive moving average model. Here the model for values $\{X_t\}$ in a time series can be written in the form

$$X_t = c + \varepsilon_t + \sum_{i=1}^{p} \varphi_i X_{t-i} + \sum_{i=1}^{q} \theta_i \varepsilon_{t-i}.$$

where again the quantities ε_t are random variables representing innovations which are new random effects that appear at a certain time but also affect values of X at later times. In this instance the use of the term "linear model" refers to the structure of the above relationship in representing X_t as a linear function of past values of the same time series and of current and past values of the innovations. This particular aspect of the structure means that it is relative simple to derive relations for the mean and covariance properties of the time series. Note that here the "linear" part of the term "linear model" is not referring to the coefficients φ_i and θ_i, as it would be in the case of a regression model, which looks structurally similar.

Other Uses in Statistics

There are some other instances where "nonlinear model" is used to contrast with a linearly structured model, although the term "linear model" is not usually applied. One example of this is nonlinear dimensionality reduction.

Matrix (Mathematics)

In mathematics, a matrix (plural matrices) is a rectangular array of numbers, symbols, or expressions, arranged in *rows* and *columns*. The individual items in a matrix are called its *elements* or *entries*. An example of a matrix with 2 rows and 3 columns is

$$\begin{bmatrix} 1 & 9 & -13 \\ 20 & 5 & -6 \end{bmatrix}.$$

Matrices of the same size can be added or subtracted element by element. The rule for matrix multiplication is more complicated, and two matrices can be multiplied only when the number of columns in the first equals the number of rows in the second. A major application of matrices is to represent linear transformations, that is, generalizations of linear functions such as $f(x) = 4x$.

For example, the rotation of vectors in three dimensional space is a linear transformation. If R is a rotation matrix and v is a column vector (a matrix with only one column) describing the position of a point in space, the product Rv is a column vector describing the position of that point after a rotation.

The product of two matrices is a matrix that represents the composition of two linear transformations. Another application of matrices is in the solution of a system of linear equations. If the matrix is square, it is possible to deduce some of its properties by computing its determinant. For example, a square matrix has an inverse if and only if its determinant is not zero. Eigenvalues and eigenvectors provide insight into the geometry of linear transformations.

Applications of matrices are found in most scientific fields. In every branch of physics, including classical mechanics, optics, electromagnetism, quantum mechanics, and quantum electrodynamics, they are used to study physical phenomena, such as the motion of rigid bodies. In computer graphics, they are used to project a 3-dimensional image onto a 2-dimensional screen. In probability theory and statistics, stochastic matrices are used to describe sets of probabilities; for instance, they are used within the PageRank algorithm that ranks the pages in a Google search. Matrix calculus generalizes classical analytical notions such as derivatives and exponentials to higher dimensions.

A major branch of numerical analysis is devoted to the development of efficient algorithms for matrix computations, a subject that is centuries old and is today an expanding area of research. Matrix decomposition methods simplify computations, both theoretically and practically. Algorithms that are tailored to the structure of particular matrix structures, e.g. sparse matrices and near-diagonal matrices, expedite computations in finite element method and other computations. Infinite matrices occur in planetary theory and in atomic theory. A simple example is the matrix representing the derivative operator, which acts on the Taylor series of a function.

Definition

A *matrix* is a rectangular arrangement of mathematical expressions that can be simply numbers.

Commonly a matrix is written as values in box brackets:

$$A = \begin{bmatrix} a_{11} & a_{12} & \cdots & a_{1n} \\ a_{21} & a_{22} & \cdots & a_{2n} \\ \vdots & \vdots & \ddots & \vdots \\ a_{m1} & a_{m2} & \cdots & a_{mn} \end{bmatrix}.$$

For example,

$$\mathrm{A} = \begin{bmatrix} 9 & 13 & 5 \\ 1 & 11 & 7 \\ 3 & 7 & 2 \\ 6 & 0 & 7 \end{bmatrix}.$$

An alternative notation uses large parentheses instead of box brackets.

The horizontal and vertical lines in a matrix are called *rows* and *columns*, respectively. The numbers in the matrix are called its *entries* or its *elements*. To specify the size of a matrix, a matrix with m rows and n columns is called an m-by-n matrix or $m \times n$ matrix, while m and n are called its *dimensions*. The above is a 4-by-3 matrix.

A matrix with one row (a $1 \times n$ matrix) is called a row vector, and a matrix with one column (an $m \times 1$ matrix) is called a column vector. Any row or column of a matrix determines a row or column vector, obtained by removing all other rows or columns respectively from the matrix. For example, the row vector for the third row of the above matrix A is

$$\begin{bmatrix} 3 & 7 & 2 \end{bmatrix}.$$

When a row or column of a matrix is interpreted as a value, this refers to the corresponding row or column vector. For instance one may say that two different rows of a matrix are equal, meaning they determine the same row vector.

In some cases the value of a row or column should be interpreted just as a sequence of values (an element of R^n if entries are real numbers) rather than as a matrix, for instance when saying that the rows of a matrix are equal to the corresponding columns of its transpose matrix.

Most of this article focuses on *real* and *complex matrices*, i.e., matrices whose elements are real or complex numbers, respectively. More general types of entries are discussed below.

Notation

The specifics of matrices notation varies widely, with some prevailing trends. Matrices are usually denoted using upper-case letters, while the corresponding lower-case letters, with two subscript indices, represent the entries. In addition to using upper-case letters to symbolize matrices, many authors use a special typographical style, commonly boldface upright (non-italic), to further distinguish matrices from other mathematical objects. An alternative notation involves the use of a double-underline with the variable name, with or without boldface style, (e.g., $\underline{\underline{A}}$).

The entry in the i-th row and the j-th column of a matrix is typically referred to as the i,j, (i,j), or $(i,j)^{\text{th}}$ entry of the matrix. For example, the (1,3) entry of the above matrix A is 5. The $(i, j)^{th}$ entry of a matrix A is most commonly written as $a_{i,j}$. Alternative notations for that entry are $A[i,j]$ or $A_{i,j}$.

Sometimes a matrix is referred to by giving a formula for its $(i,j)^{th}$ entry, often with double parenthesis around the formula for the entry, for example, if the $(i,j)^{th}$ entry of A were given by a_{ij}, A would be denoted $((a_{ij}))$.

An asterisk is occasionally used to refer to whole rows or columns in a matrix. For example, $a_{i,*}$ refers to the ith row of A, and $a_{*,j}$ refers to the jth column of A. The set of all m-by-n matrices is denoted $\mathbb{M}(m, n)$.

A common shorthand is

$$A = [a_{i,j}]_{i = 1,\ldots,m;\ j = 1,\ldots,n} \text{ or more briefly } A = [a_{i,j}]_{m\times n}$$

to define an $m \times n$ matrix A. Usually the entries $a_{i,j}$ are defined separately for all integers $1 \le i \le m$ and $1 \le j \le n$. They can however sometimes be given by one formula; for example the 3-by-4 matrix

$$A = \begin{bmatrix} 0 & -1 & -2 & -3 \\ 1 & 0 & -1 & -2 \\ 2 & 1 & 0 & -1 \end{bmatrix}$$

can alternatively be specified by $A = [i - j]_{i = 1,2,3;\ j = 1,\ldots,4}$, or simply $A = ((i-j))$, where the size of the matrix is understood.

Some programming languages start the numbering of rows and columns at zero, in which case the entries of an m-by-n matrix are indexed by $0 \le i \le m - 1$ and $0 \le j \le n - 1$. This article follows the more common convention in mathematical writing where enumeration starts from 1.

Matrix Multiplication, Linear Equations and Linear Transformations

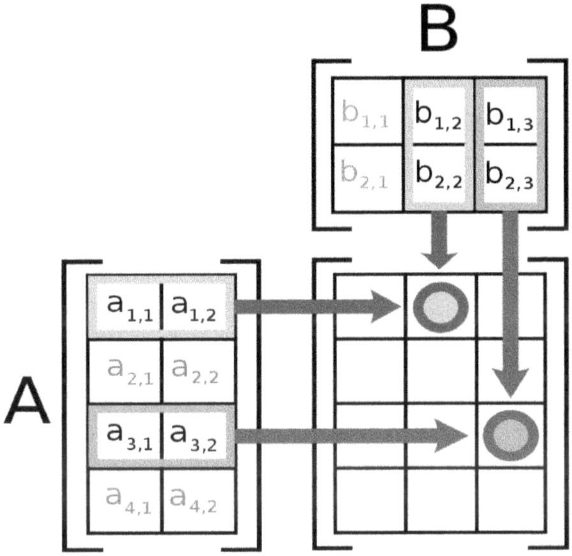

Figure: *Schematic depiction of the matrix product AB of two matrices A and B.*

Multiplication of two matrices is defined only if the number of columns of the left matrix is the same as the number of rows of the right matrix. If A is an m-by-n matrix and B is an n-by-p matrix, then their *matrix product AB* is the m-by-p matrix whose entries are given by dot product of the corresponding row of A and the corresponding column of B:

$$[AB]_{i,j} = A_{i,1}B_{1,j} + A_{i,2}B_{2,j} + \cdots + A_{i,n}B_{n,j} = \sum_{r=1}^{n} A_{i,r}B_{r,j} ,$$

where $1 \le i \le m$ and $1 \le j \le p$. For example, the underlined entry 2340 in the product is calculated as $(2 \times 1000) + (3 \times 100) + (4 \times 10) = 2340$:

$$\begin{bmatrix} \underline{2} & \underline{3} & \underline{4} \\ 1 & 0 & 0 \end{bmatrix} \begin{bmatrix} 0 & \underline{1000} \\ 1 & \underline{100} \\ 0 & \underline{10} \end{bmatrix} = \begin{bmatrix} 3 & \underline{2340} \\ 0 & 1000 \end{bmatrix}.$$

Matrix multiplication satisfies the rules $(AB)C = A(BC)$ (associativity), and $(A+B)C = AC+BC$ as well as $C(A+B) = CA+CB$ (left and right distributivity), whenever the size of the matrices is such that the various products are defined.

The product AB may be defined without BA being defined, namely if A and B are m-by-n and n-by-k matrices, respectively, and $m \neq k$. Even if both products are defined, they need not be equal, i.e., generally one has

$$AB \neq BA,$$

i.e., *matrix multiplication is not commutative,* in marked contrast to (rational, real, or complex) numbers whose product is independent of the order of the factors. An example of two matrices not commuting with each other is:

$$\begin{bmatrix} 1 & 2 \\ 3 & 4 \end{bmatrix} \begin{bmatrix} 0 & 1 \\ 0 & 0 \end{bmatrix} = \begin{bmatrix} 0 & 1 \\ 0 & 3 \end{bmatrix},$$

whereas

$$\begin{bmatrix} 0 & 1 \\ 0 & 0 \end{bmatrix} \begin{bmatrix} 1 & 2 \\ 3 & 4 \end{bmatrix} = \begin{bmatrix} 3 & 4 \\ 0 & 0 \end{bmatrix}.$$

The identity matrix I_n of size n is the n-by-n matrix in which all the elements on the main diagonal are equal to 1 and all other elements are equal to 0, e.g.

$$I_3 = \begin{bmatrix} 1 & 0 & 0 \\ 0 & 1 & 0 \\ 0 & 0 & 1 \end{bmatrix}.$$

It is called identity matrix because multiplication with it leaves a matrix unchanged: $MI_n = I_mM = M$ for any m-by-n matrix M.

Besides the ordinary matrix multiplication just described, there exist other less frequently used operations on matrices that can be considered forms of multiplication, such as the Hadamard product and the Kronecker product. They arise in solving matrix equations such as the Sylvester equation.

Linear Equations

Matrices can be used to compactly write and work with multiple linear equations, i.e., systems of linear equations. For example, if A is an m-by-n matrix, x designates a column vector (i.e., $n\times1$-matrix) of n variables x_1, x_2, ..., x_n, and b is an $m\times1$-column vector, then the matrix equation

$$Ax = b$$

is equivalent to the system of linear equations

$$A_{1,1}x_1 + A_{1,2}x_2 + \dots + A_{1,n}x_n = b_1$$

$$\dots$$

$$A_{m,1}x_1 + A_{m,2}x_2 + \dots + A_{m,n}x_n = b_m \ .$$

Square Matrices

A *square matrix* is a matrix with the same number of rows and columns. An n-by-n matrix is known as a square matrix of order n. Any two square matrices of the same order can be added and multiplied. A square matrix A is called *invertible* or *non-singular* if there exists a matrix B such that

$$AB = I_n.$$

This is equivalent to $BA = I_n$. Moreover, if B exists, it is unique and is called the *inverse matrix* of A, denoted A^{-1}.

The entries $A_{i,i}$ form the main diagonal of a matrix. The trace, tr(A) of a square matrix A is the sum of its diagonal entries. While matrix multiplication is not commutative as mentioned above, the trace of the product of two matrices is independent of the order of the factors: tr(AB) = tr(BA).

Also, the trace of a matrix is equal to that of its transpose, i.e., tr(A) = tr(A^{T}).

If all entries outside the main diagonal are zero, A is called a diagonal matrix. If only all entries above (below) the main diagonal are zero, A is called a lower triangular matrix (upper triangular matrix, respectively). For example, if $n = 3$, they look like

$$\begin{bmatrix} d_{11} & 0 & 0 \\ 0 & d_{22} & 0 \\ 0 & 0 & d_{33} \end{bmatrix} \text{(diagonal)}, \begin{bmatrix} l_{11} & 0 & 0 \\ l_{21} & l_{22} & 0 \\ l_{31} & l_{32} & l_{33} \end{bmatrix} \text{(lower) and } \begin{bmatrix} u_{11} & u_{12} & u_{13} \\ 0 & u_{22} & u_{23} \\ 0 & 0 & u_{33} \end{bmatrix} \text{(upper}$$

triangular matrix).

Determinant

The *determinant* det(A) or $|A|$ of a square matrix A is a number encoding certain properties of the matrix. A matrix is invertible if and only if its determinant is nonzero.

Its absolute value equals the area (in R^2) or volume (in R^3) of the image of the unit square (or cube), while its sign corresponds to the orientation of the corresponding linear map: the determinant is positive if and only if the orientation is preserved.

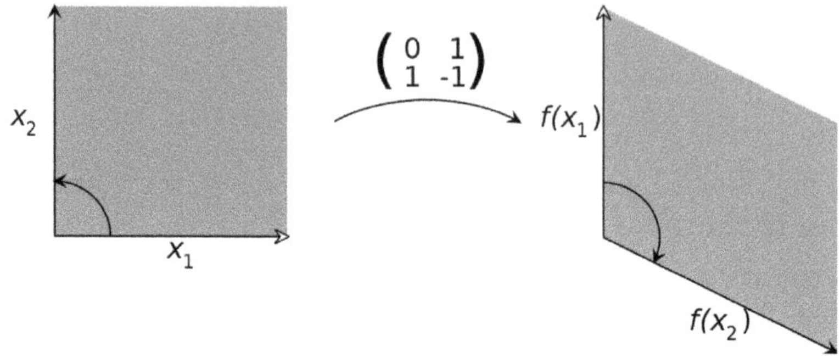

***Figure:** A linear transformation on R^2 given by the indicated matrix. The determinant of this matrix is –1, as the area of the green parallelogram at the right is 1, but the map reverses the orientation, since it turns the counterclockwise orientation of the vectors to a clockwise one.*

The determinant of 2-by-2 matrices is given by

$$\det\begin{bmatrix} a & b \\ c & d \end{bmatrix} = ad - bc.$$

When the determinant is equal to one, then the matrix represents an equi-areal mapping. The determinant of 3-by-3 matrices involves 6 terms (rule of Sarrus). The more lengthy Leibniz formula generalises these two formulae to all dimensions.

The determinant of a product of square matrices equals the product of their determinants: $\det(AB) = \det(A) \cdot \det(B)$. Adding a multiple of any row to another row, or a multiple of any column to another column, does not change the determinant.

Interchanging two rows or two columns affects the determinant by multiplying it by –1. Using these operations, any matrix can be transformed to a lower (or upper) triangular matrix, and for such matrices the determinant equals the product of the entries on the main diagonal; this provides a method to calculate the determinant of any matrix.

Finally, the Laplace expansion expresses the determinant in terms of minors, i.e., determinants of smaller matrices. This expansion can be used for a recursive definition of determinants (taking as starting case the determinant of a 1-by-1 matrix, which is its unique entry, or even the determinant of a 0-by-0 matrix, which is 1), that can be seen to be equivalent to the Leibniz formula. Determinants can be used to solve linear systems using Cramer's rule, where the division of the determinants of two related square matrices equates to the value of each of the system's variables.

Eigenvalues and Eigenvectors

A number λ and a non-zero vector v satisfying

$$Av = \lambda v$$

are called an *eigenvalue* and an *eigenvector* of A, respectively. The number λ is an eigenvalue of an $n \times n$-matrix A if and only if $A - \lambda I_n$ is not invertible, which is equivalent to

$$\det(A - \lambda I) = 0.$$

The polynomial p_A in an indeterminate X given by evaluation the determinant $\det(XI_n - A)$ is called the characteristic polynomial of A. It is a monic polynomial of degree n. Therefore the polynomial equation $p_A(\lambda) = 0$ has at most n different solutions, i.e., eigenvalues of the matrix. They may be complex even if the entries of A are real. According to the Cayley–Hamilton theorem, $p_A(A) = 0$, that is, the result of substituting the matrix itself into its own characteristic polynomial yields the zero matrix.

Symmetry

A square matrix A that is equal to its transpose, i.e., $A = A^T$, is a symmetric matrix. If instead, A was equal to the negative of its transpose, i.e., $A = -A^T$, then A is a skew-symmetric matrix. In complex matrices, symmetry is often replaced by the concept of Hermitian matrices, which satisfy $A^- = A$, where the star or asterisk denotes the conjugate transpose of the matrix, i.e., the transpose of the complex conjugate of A.

By the spectral theorem, real symmetric matrices and complex Hermitian matrices have an eigenbasis; i.e., every vector is expressible as a linear combination of eigenvectors. In both cases, all eigenvalues are real. This theorem can be generalized to infinite-dimensional situations related to matrices with infinitely many rows and columns.

Computational Aspects

In addition to theoretical knowledge of properties of matrices and their relation to other fields, it is important for practical purposes to perform matrix calculations effectively and precisely. The domain studying these matters is called numerical linear algebra. As with other numerical situations, two main aspects are the complexity of algorithms and their numerical stability. Many problems can be solved by both direct algorithms or iterative approaches. For example, finding eigenvectors can be done by finding a sequence of vectors x_n converging to an eigenvector when n tends to infinity.

Determining the complexity of an algorithm means finding upper bounds or estimates of how many elementary operations such as additions and multiplications of scalars are necessary to perform some algorithm, e.g., multiplication of matrices. For example, calculating the matrix product of two n-by-n matrix using the definition given above needs n^3 multiplications, since for any of the n^2 entries of the product, n multiplications are necessary. The Strassen algorithm outperforms this "naive" algorithm; it needs only $n^{2.807}$ multiplications. A refined approach also incorporates specific features of the computing devices.

In many practical situations additional information about the matrices involved is known. An important case are sparse matrices, i.e., matrices most of whose entries are zero. There are specifically adapted algorithms for, say, solving linear systems $Ax = b$ for sparse matrices A, such as the conjugate gradient method.

An algorithm is, roughly speaking, numerically stable, if little deviations (such as rounding errors) do not lead to big deviations in the result. For example, calculating the inverse of a matrix via Laplace's formula (Adj (A) denotes the adjugate matrix of A)

$$A^{-1} = \text{Adj}(A) \;/\; \det(A)$$

may lead to significant rounding errors if the determinant of the matrix is very small. The norm of a matrix can be used to capture the conditioning of linear algebraic problems, such as computing a matrix' inverse.

Although most computer languages are not designed with commands or libraries for matrices, as early as the 1970s, some engineering desktop computers such as the HP 9830 had ROM cartridges to add BASIC commands for matrices. Some computer languages such as APL were designed to manipulate matrices, and various mathematical programmes can be used to aid computing with matrices.

Matrix Decomposition Methods

There are several methods to render matrices into a more easily accessible form. They are generally referred to as *matrix transformation* or *matrix decomposition* techniques. The interest of all these techniques is that they preserve certain properties of the matrices in question, such as determinant, rank or inverse, so that these quantities can be calculated after applying the transformation, or that certain matrix operations are algorithmically easier to carry out for some types of matrices.

The LU decomposition factors matrices as a product of lower (L) and an upper triangular matrices (U). Once this decomposition is calculated, linear systems can be solved more efficiently, by a simple technique called forward and back substitution. Likewise, inverses of triangular matrices are algorithmically easier to calculate.

The *Gaussian elimination* is a similar algorithm; it transforms any matrix to row echelon form. Both methods proceed by multiplying the matrix by suitable elementary matrices, which correspond to permuting rows or columns and adding multiples of one row to another row. Singular value decomposition expresses any matrix *A* as a product UDV^{-}, where *U* and *V* are unitary matrices and *D* is a diagonal matrix.

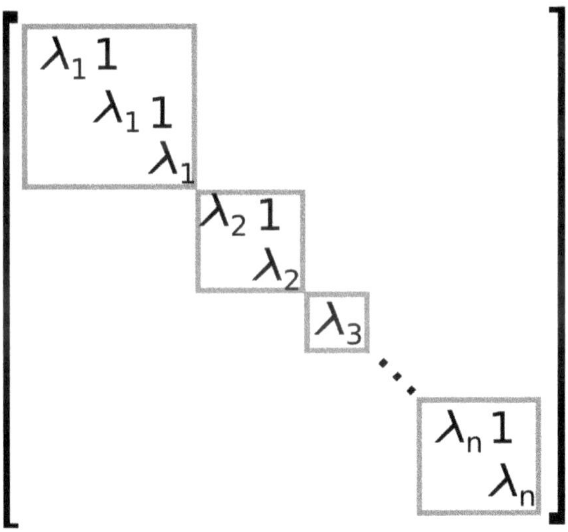

Figure: *An example of a matrix in Jordan normal form. The grey blocks are called Jordan blocks.*

The eigendecomposition or *diagonalization* expresses *A* as a product VDV^{-1}, where *D* is a diagonal matrix and *V* is a suitable invertible matrix. If *A* can be written in this form, it is called diagonalizable.

More generally, and applicable to all matrices, the Jordan decomposition transforms a matrix into Jordan normal form, that is to say matrices whose only nonzero entries are the eigenvalues λ_1 to λ_n of *A*, placed on the main diagonal and possibly entries equal to one directly above the main diagonal, as shown at the right. Given the eigende composition, the nth power of *A* (i.e., *n*-fold iterated matrix multiplication) can be calculated via

$$A^n = (VDV^{-1})^n = VDV^{-1}VDV^{-1}...VDV^{-1} = VD^nV^{-1}$$

and the power of a diagonal matrix can be calculated by taking the corresponding powers of the diagonal entries, which is much easier than doing the exponentiation for A instead. This can be used to compute the matrix exponential e^A, a need frequently arising in solving linear differential equations, matrix logarithms and square roots of matrices. To avoid numerically ill-conditioned situations, further algorithms such as the Schur decomposition can be employed.

Abstract Algebraic Aspects and Generalizations

Matrices can be generalized in different ways. Abstract algebra uses matrices with entries in more general fields or even rings, while linear algebra codifies properties of matrices in the notion of linear maps. It is possible to consider matrices with infinitely many columns and rows. Another extension are tensors, which can be seen as higher-dimensional arrays of numbers, as opposed to vectors, which can often be realised as sequences of numbers, while matrices are rectangular or two-dimensional array of numbers. Matrices, subject to certain requirements tend to form groups known as matrix groups.

Matrices with More General Entries

This article focuses on matrices whose entries are real or complex numbers. *However, matrices can be considered with much more general types of entries than real or complex numbers.* As a first step of generalization, any field, i.e., a set where addition, subtraction, multiplication and division operations are defined and well-behaved, may be used instead of R or C, for example rational numbers or finite fields. For example, coding theory makes use of matrices over finite fields. Wherever eigenvalues are considered, as these are roots of a polynomial they may exist only in a larger field than that of the coefficients of the matrix; for instance they may be complex in case of a matrix with real entries. The possibility to reinterpret the entries of a matrix as elements of a larger field (e.g., to view a real matrix as a complex matrix whose entries happen to be all real) then allows considering each square matrix to possess a full set of eigenvalues. Alternatively one can consider only matrices with entries in an algebraically closed field, such as C, from the outset.

More generally, abstract algebra makes great use of matrices with entries in a ring R. Rings are a more general notion than fields in that no division operation exists. The very same addition and multiplication operations of matrices extend to this setting, too. The set $M(n, R)$ of all square n-by-n matrices over R is a ring called matrix ring, isomorphic to the endomorphism ring of the left R-moduleR^n. If the ring R is commutative, i.e., its multiplication is commutative, then M(n, R) is a unitary noncommutative (unless $n = 1$) associative algebra over R. The determinant of square matrices over a commutative ring R can still be defined using the Leibniz formula; such a matrix is invertible if and only if its determinant is invertible in R, generalising the situation over a field F, where every nonzero element is invertible. Matrices over superrings are called supermatrices.

Matrices do not always have all their entries in the same ring – or even in any ring at all. One special but common case is block matrices, which may be considered as matrices whose entries themselves are matrices. The entries need not be quadratic matrices, and thus need not be members of any ordinary ring; but their sizes must fulfil certain compatibility conditions.

Relationship to Linear Maps

Linear maps $R^n \rightarrow R^m$ are equivalent to m-by-n matrices, as described above. More generally, any linear map f: $V \rightarrow W$ between finite-dimensionalvector spaces can be described by a matrix $A = (a_{ij})$, after choosing bases $v_1, ..., v_n$ of V, and $w_1, ..., w_m$ of W (so n is the dimension of V and m is the dimension of W), which is such that

$$f(\mathrm{v}_j) = \sum_{i=1}^{m} a_{i,j} \mathrm{w}_i \qquad \text{for } j = 1,\ldots,n.$$

In other words, column j of A expresses the image of v_j in terms of the basis vectors w_i of W; thus this relation uniquely determines the entries of the matrix A. Note that the matrix depends on the choice of the bases: different choices of bases give rise to different, but equivalent matrices. Many of the above concrete notions can be reinterpreted in this light, for example, the transpose matrix A^T describes the transpose of the linear map given by A, with respect to the dual bases.

More generally, the set of $m\times n$ matrices can be used to represent the R-linear maps between the free modules R^m and R^n for an arbitrary ring R with unity. When $n = m$ composition of these maps is possible, and this gives rise to the matrix ring of $n\times n$ matrices representing the endomorphism ring of R^n.

Matrix Groups

A group is a mathematical structure consisting of a set of objects together with a binary operation, i.e., an operation combining any two objects to a third, subject to certain requirements. A group in which the objects are matrices and the group operation is matrix multiplication is called a *matrix group*. Since in a group every element has to be invertible, the most general matrix groups are the groups of all invertible matrices of a given size, called the general linear groups.

Any property of matrices that is preserved under matrix products and inverses can be used to define further matrix groups. For example, matrices with a given size and with a determinant of 1 form a subgroup of (i.e., a smaller group contained in) their general linear group, called a special linear group. Orthogonal matrices, determined by the condition

$$M^T M = I,$$

form the orthogonal group. They are called *orthogonal* since the associated linear transformations of R^n preserve angles in the sense that the scalar product of two vectors is unchanged after applying M to them:

$$(Mv) \cdot (Mw) = v \cdot w.$$

Every finite group is isomorphic to a matrix group, as one can see by considering the regular representation of the symmetric group. General groups can be studied using matrix groups, which are comparatively well-understood, by means of representation theory.

Infinite Matrices

It is also possible to consider matrices with infinitely many rows and/or columns even if, being infinite objects, one cannot write down such matrices explicitly. All that matters is that for every element in the set indexing rows, and every element in the set indexing columns, there is a well-defined entry (these index sets need not even be subsets of the natural numbers). The basic operations of addition, subtraction, scalar multiplication and transposition can still be defined without problem; however matrix multiplication may involve infinite summations to define the resulting entries, and these are not defined in general.

If R is any ring with unity, then the ring of endomorphisms of $M = \bigoplus_{i \in I} R$ as a right R module is isomorphic to the ring of column finite matrices $\mathbb{CFM}_I(R)$ whose entries are indexed by $I \times I$, and whose columns each contain only finitely many nonzero entries. The endomorphisms of M considered as a left R module result in an analogous object, the row finite matrices $\mathbb{RFM}_I(R)$ whose rows each only have finitely many nonzero entries.

If infinite matrices are used to describe linear maps, then only those matrices can be used all of whose columns have but a finite number of nonzero entries, for the following reason. For a matrix A to describe a linear map f: $V \to W$, bases for both spaces must have been chosen; recall that by definition this means that every vector in the space can be written uniquely as a (finite) linear combination of basis vectors, so that written as a (column) vector v of coefficients, only finitely many entries v_i are nonzero. Now the columns of A describe the images by f of individual basis vectors of V in the basis of W, which is only meaningful if these columns have only finitely many nonzero entries.

There is no restriction on the rows of A however: in the product $A \cdot v$ there are only finitely many nonzero coefficients of v involved, so every one of its entries, even if it is given as an infinite sum of products, involves only finitely many nonzero terms and is therefore well defined. Moreover this amounts to forming a linear combination of the columns of A that effectively involves only finitely many of them, whence the result has only finitely many nonzero entries, because each of those columns do. One also sees that products of two matrices of the given type is well defined (provided as usual that the column-index and row-index sets match), is again of the same type, and corresponds to the composition of linear maps.

If R is a normed ring, then the condition of row or column finiteness can be relaxed. With the norm in place, absolutely convergent series can be used instead of finite sums. For example, the matrices whose column sums are absolutely convergent sequences form a ring. Analogously of course, the matrices whose row sums are absolutely convergent series also form a ring.

In that vein, infinite matrices can also be used to describe operators on Hilbert spaces, where convergence and continuity questions arise, which again results in certain constraints that have to be imposed. However, the explicit point of view of matrices tends to obfuscate the matter, and the abstract and more powerful tools of functional analysis can be used instead.

Empty Matrices

An *empty matrix* is a matrix in which the number of rows or columns (or both) is zero. Empty matrices help dealing with maps involving the zero vector space. For example, if A is a 3-by-0 matrix and B is a 0-by-3 matrix, then AB is the 3-by-3 zero matrix corresponding to the null map from a 3-dimensional space V to itself, while *BA* is a 0-by-0 matrix. There is no common notation for empty matrices, but most computer algebra systems allow creating and computing with them. The determinant of the 0-by-0 matrix is 1 as follows from regarding the empty product occurring in the Leibniz formula for the determinant as 1. This value is also consistent with the fact that the identity map from any finite dimensional space to itself has determinant 1, a fact that is often used as a part of the characterization of determinants.

Applications

There are numerous applications of matrices, both in mathematics and other sciences. Some of them merely take advantage of the compact representation of a set of numbers in a matrix. For example, in game theory and economics, the payoff matrix encodes the payoff for two players, depending on which out of a given (finite) set of alternatives the players choose. Text mining and automated thesaurus compilation makes use of document-term matrices such as tf-idf to track frequencies of certain words in several documents.

Complex numbers can be represented by particular real 2-by-2 matrices via

$$a+ib \leftrightarrow \begin{bmatrix} a & -b \\ b & a \end{bmatrix},$$

under which addition and multiplication of complex numbers and matrices correspond to each other. For example, 2-by-2 rotation matrices represent the multiplication with some complex number of absolute value 1, as above. A similar interpretation is possible for quaternions, and also for Clifford algebras in general.

Early encryption techniques such as the Hill cipher also used matrices. However, due to the linear nature of matrices, these codes are comparatively easy to break. Computer graphics uses matrices both to represent objects and to calculate transformations of objects using affine rotation matrices to accomplish tasks such as projecting a three-dimensional object onto a two-dimensional screen, corresponding to a theoretical camera observation. Matrices over a polynomial ring are important in the study of control theory.

Chemistry makes use of matrices in various ways, particularly since the use of quantum theory to discuss molecular bonding and spectroscopy. Examples are the overlap matrix and the Fock matrix used in solving the Roothaan equations to obtain the molecular orbitals of the Hartree–Fock method.

Graph Theory

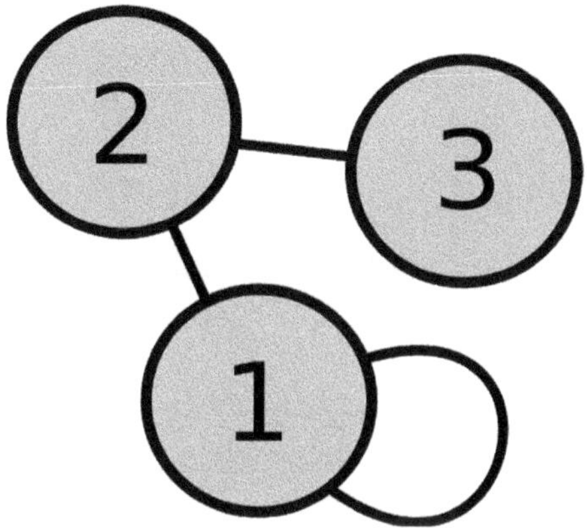

***Figure:** An undirected graph with adjacency matrix* $\begin{bmatrix} 1 & 1 & 0 \\ 1 & 0 & 1 \\ 0 & 1 & 0 \end{bmatrix}$.

The adjacency matrix of a finite graph is a basic notion of graph theory. It saves which vertices of the graph are connected by an edge. Matrices containing just two different values (0 and 1 meaning for example "yes" and "no") are called logical matrices. The distance (or cost) matrix contains information about distances of the edges. These concepts can be applied to websites connected hyperlinks or cities connected by roads etc., in which case (unless the road network is extremely dense) the matrices tend to be sparse, i.e., contain few nonzero entries. Therefore, specifically tailored matrix algorithms can be used in network theory.

Analysis and Geometry

The Hessian matrix of a differentiable function f: $R^n \to R$ consists of the second derivatives of f with respect to the several coordinate directions, i.e.

$$H(f) = \left[\frac{\partial^2 f}{\partial x_i \partial x_j} \right].$$

It encodes information about the local growth behaviour of the function: given a critical point $x = (x_1, ..., x_n)$, i.e., a point where the first partial derivatives $\partial f / \partial x_i$ of f vanish, the function has a local minimum if the Hessian matrix is positive definite. Quadratic programming can be used to find global minima or maxima of quadratic functions closely related to the ones attached to matrices.

Another matrix frequently used in geometrical situations is the *Jacobi matrix* of a differentiable map $f: R^n \rightarrow R^m$. If $f_1, ..., f_m$ denote the components of f, then the Jacobi matrix is defined as.

$$J(f) = \left[\frac{\partial f_i}{\partial x_j}\right]_{1 \le i \le m, 1 \le j \le n}.$$

If $n>m$, and if the rank of the Jacobi matrix attains its maximal value m, f is locally invertible at that point, by the implicit function theorem.

Partial differential equations can be classified by considering the matrix of coefficients of the highest-order differential operators of the equation. For elliptic partial differential equations this matrix is positive definite, which has decisive influence on the set of possible solutions of the equation in question.

The finite element method is an important numerical method to solve partial differential equations, widely applied in simulating complex physical systems.

It attempts to approximate the solution to some equation by piecewise linear functions, where the pieces are chosen with respect to a sufficiently fine grid, which in turn can be recast as a matrix equation.

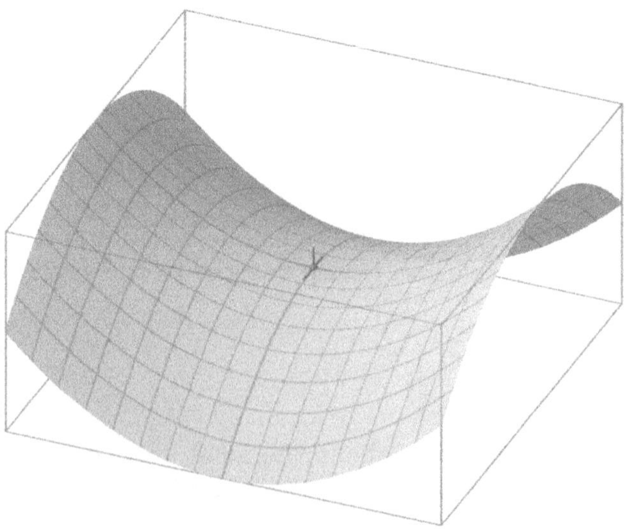

Figure: *At the saddle point ($x = 0$, $y = 0$) (red) of the function $f(x,-y) = x^2 - y^2$, the Hessian matrix* $\begin{bmatrix} 2 & 0 \\ 0 & -2 \end{bmatrix}$ *is indefinite.*

Probability Theory and Statistics

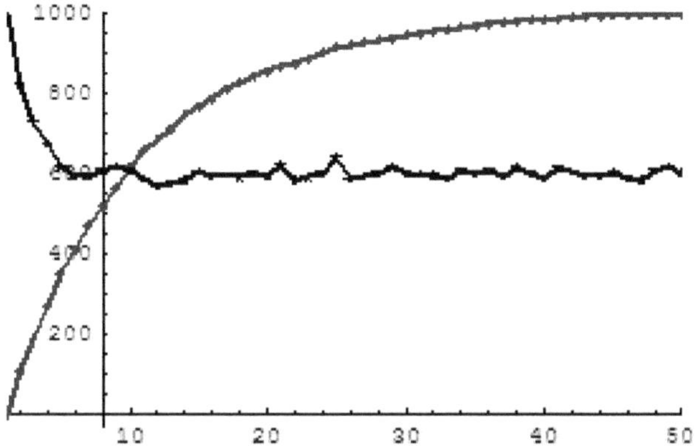

Figure: *Two different Markov chains. The chart depicts the number of particles (of a total of 1000) in state –2". Both limiting values can be determined from the transition matrices, which are given by* $\begin{bmatrix} .7 & 0 \\ .3 & 1 \end{bmatrix}$ *(red) and* $\begin{bmatrix} .7 & .2 \\ .3 & .8 \end{bmatrix}$ *(black).*

Stochastic matrices are square matrices whose rows are probability vectors, i.e., whose entries are non-negative and sum up to one. Stochastic matrices are used to define Markov chains with finitely many states. A row of the stochastic matrix gives the probability distribution for the next position of some particle currently in the state that corresponds to the row. Properties of the Markov chain like absorbing states, i.e., states that any particle attains eventually, can be read off the eigenvectors of the transition matrices.

Statistics also makes use of matrices in many different forms. Descriptive statistics is concerned with describing data sets, which can often be represented in matrix form, by reducing the amount of data. The covariance matrix encodes the mutual variance of several random variables. Another technique using matrices are linear least squares, a method that approximates a finite set of pairs (x_1, y_1), (x_2, y_2), ..., (x_N, y_N), by a linear function

$$y_i \approx ax_i + b, \; i = 1, \ldots, N$$

which can be formulated in terms of matrices, related to the singular value decomposition of matrices.

Random matrices are matrices whose entries are random numbers, subject to suitable probability distributions, such as matrix normal distribution. Beyond probability theory, they are applied in domains ranging from number theory to physics.

Symmetries and Transformations in Physics

Linear transformations and the associated symmetries play a key role in modern physics. For example, elementary particles in quantum field theory are classified as representations of the Lorentz group of special relativity and, more specifically, by their behaviour under the spin group. Concrete representations involving the Pauli matrices and more general gamma matrices are an integral part of the physical description of fermions, which behave as spinors. For the three lightest quarks, there is a group-theoretical representation involving the special unitary group SU(3); for their calculations, physicists use a convenient matrix representation known as the Gell-Mann matrices, which are also used for the SU(3) gauge group that forms the basis of the modern description of strong nuclear interactions, quantum chromodynamics. The Cabibbo–Kobayashi–Maskawa matrix, in turn, expresses the fact that the basic quark states that are important for weak interactions are not the same as, but linearly related to the basic quark states that define particles with specific and distinct masses.

Linear Combinations of Quantum States

The first model of quantum mechanics (Heisenberg, 1925) represented the theory's operators by infinite-dimensional matrices acting on quantum states. This is also referred to as matrix mechanics. One particular example is the density matrix that characterizes the "mixed" state of a quantum system as a linear combination of elementary, "pure" eigenstates.

Another matrix serves as a key tool for describing the scattering experiments that form the cornerstone of experimental particle physics: Collision reactions such as occur in particle accelerators, where non-interacting particles head towards each other and collide in a small interaction zone, with a new set of non-interacting particles as the result, can be described as the scalar product of outgoing particle states and a linear combination of ingoing particle states. The linear combination is given by a matrix known as the S-matrix, which encodes all information about the possible interactions between particles.

Normal Modes

A general application of matrices in physics is to the description of linearly coupled harmonic systems. The equations of motion of such systems can be described in matrix form, with a mass matrix multiplying a generalized velocity to give the kinetic term, and a force matrix multiplying a displacement vector to characterize the interactions.

The best way to obtain solutions is to determine the system's eigenvectors, its normal modes, by diagonalizing the matrix equation. Techniques like this are crucial when it comes to the internal dynamics of molecules: the internal vibrations of systems consisting of mutually bound component atoms. They are also needed for describing mechanical vibrations, and oscillations in electrical circuits.

Geometrical Optics

Geometrical optics provides further matrix applications. In this approximative theory, the wave nature of light is neglected. The result is a model in which light rays are indeed geometrical rays. If the deflection of light rays by optical elements is small, the action of a lens or reflective element on a given light ray can be expressed as multiplication of a two-component vector with a two-by-two matrix called ray transfer matrix: the vector's components are the light ray's slope and its distance from the optical axis, while the matrix encodes the properties of the optical element.

Actually, there are two kinds of matrices, viz. a *refraction matrix* describing the refraction at a lens surface, and a *translation matrix*, describing the translation of the plane of reference to the next refracting surface, where another refraction matrix applies. The optical system, consisting of a combination of lenses and/or reflective elements, is simply described by the matrix resulting from the product of the components' matrices.

Electronics

Traditional mesh analysis in electronics leads to a system of linear equations that can be described with a matrix.

The behaviour of many electronic components can be described using matrices. Let A be a 2-dimensional vector with the component's input voltage v_1 and input current i_1 as its elements, and let B be a 2-dimensional vector with the component's output voltage v_2 and output current i_2 as its elements. Then the behaviour of the electronic component can be described by $B = H \cdot A$, where H is a 2 x 2 matrix containing one impedance element (h_{12}), one admittance element (h_{21}) and two dimensionless elements (h_{11} and h_{22}). Calculating a circuit now reduces to multiplying matrices.

History

Matrices have a long history of application in solving linear equations. The Chinese text *The Nine Chapters on the Mathematical Art* (*Jiu Zhang Suan Shu*), from between 300 BC and AD 200, is the first example of the use of matrix methods to solve simultaneous equations, including the concept of determinants, over 1000 years before its publication by the Japanese mathematician Seki in 1683 and the German mathematician Leibniz in 1693. Cramer presented his rule in 1750.

Early matrix theory emphasized determinants more strongly than matrices and an independent matrix concept akin to the modern notion emerged only in 1858, with Cayley's *Memoir on the theory of matrices*. The term "matrix" (Latin for "womb", derived from *mater*—mother) was coined by Sylvester in 1850, who understood a matrix as an object giving rise to a number of determinants today called minors, that is to say, determinants of smaller matrices that derive from the original one by removing columns and rows. In an 1851 paper, Sylvester explains:

I have in previous papers defined a "Matrix" as a rectangular array of terms, out of which different systems of determinants may be engendered as from the womb of a common parent.

The study of determinants sprang from several sources. Number-theoretical problems led Gauss to relate coefficients of quadratic forms, i.e., expressions such as $x^2 + xy - 2y^2$, and linear maps in three dimensions to matrices. Eisenstein further developed these notions, including the remark that, in modern parlance, matrix products are non-commutative. Cauchy was the first to prove general statements about determinants, using as definition of the determinant of a matrix $A = [a_{i,j}]$ the following: replace the powers a_j^k by a_{jk} in the polynomial

$$a_1 a_2 \cdots a_n \prod_{i<j} (a_j - a_i) \quad ,$$

where Π denotes the product of the indicated terms. He also showed, in 1829, that the eigenvalues of symmetric matrices are real. Jacobi studied "functional determinants"—later called Jacobi determinants by Sylvester—which can be used to describe geometric transformations at a local (or infinitesimal) level, Kronecker's *Vorlesungen über die Theorie der Determinanten* and Weierstrass' *Zur Determinantentheorie*, both published in 1903, first treated determinants axiomatically, as opposed to previous more concrete approaches such as the mentioned formula of Cauchy. At that point, determinants were firmly established.

Many theorems were first established for small matrices only, for example the Cayley–Hamilton theorem was proved for 2×2 matrices by Cayley in the aforementioned memoir, and by Hamilton for 4×4 matrices. Frobenius, working on bilinear forms, generalized the theorem to all dimensions (1898).

Also at the end of the 19th century the Gauss–Jordan elimination (generalizing a special case now known as Gauss elimination) was established by Jordan. In the early 20th century, matrices attained a central role in linear algebra. Partially due to their use in classification of the hypercomplex number systems of the previous century.

The inception of matrix mechanics by Heisenberg, Born and Jordan led to studying matrices with infinitely many rows and columns. Later, von Neumann carried out the mathematical formulation of quantum mechanics, by further developing functional analytic notions such as linear operators on Hilbert spaces, which, very roughly speaking, correspond to Euclidean space, but with an infinity of independent directions.

Other Historical Usages of the word "Matrix" in Mathematics

The word has been used in unusual ways by at least two authors of historical importance.

Bertrand Russell and Alfred North Whitehead in their *Principia Mathematica* (1910–1913) use the word "matrix" in the context of their Axiom of reducibility. They proposed this axiom as a means to reduce any function to one of lower type, successively, so that at the "bottom" (0 order) the function is identical to its extension:

"Let us give the name of *matrix* to any function, of however many variables, which does not involve any apparent variables. Then any possible function other than a matrix is derived from a matrix by means of generalization, i.e., by considering the proposition which asserts that the function in question is true with all possible values or with some value of one of the arguments, the other argument or arguments remaining undetermined".

For example a function $\Phi(x, y)$ of two variables x and y can be reduced to a *collection* of functions of a single variable, e.g., y, by "considering" the function for all possible values of "individuals" a_i substituted in place of variable x.

And then the resulting collection of functions of the single variable y, i.e., $\forall a_i$: $\Phi(a_i, y)$, can be reduced to a "matrix" of values by "considering" the function for all possible values of "individuals" b_i substituted in place of variable *y*:

$$\forall b_j \forall a_i: \Phi(a_i, b_j).$$

Alfred Tarski in his 1946 *Introduction to Logic* used the word "matrix" synonymously with the notion of truth table as used in mathematical logic.

Invertible Matrix

In linear algebra an *n*-by-*n* (square) matrix *A* is called invertible (some authors use nonsingular or nondegenerate) if there exists an *n*-by-*n* matrix *B* such that

$$AB = BA = I_n$$

where I_n denotes the *n*-by-*n* identity matrix and the multiplication used is ordinary matrix multiplication. If this is the case, then the matrix *B* is uniquely determined by A and is called the *inverse* of *A*, denoted by A^{-1}. It follows from the theory of matrices that if

$$AB = I$$

for *finite square* matrices *A* and *B*, then also

$$BA = I.$$

Non-square matrices (*m*-by-*n* matrices for which $m \neq n$) do not have an inverse. However, in some cases such a matrix may have a left inverse or right inverse. If A is *m*-by-*n* and the rank of *A* is equal to *n*, then A has a left inverse: an *n*-by-*m* matrix *B* such that $BA = I$. If A has rank *m*, then it has a right inverse: an *n*-by-*m* matrix *B* such that $AB = I$.

A square matrix that is not invertible is called singular or degenerate. A square matrix is singular if and only if its determinant is 0. Singular matrices are rare in the sense that if you pick a random square matrix over a continuous uniform distribution on its entries, it will almost surely not be singular.

While the most common case is that of matrices over the real or complex numbers, all these definitions can be given for matrices over any commutative ring. However, in this case the condition for a square matrix to be invertible is that its determinant is invertible in the ring, which in general is a much stricter requirement than being nonzero. The conditions for existence of left-inverse resp. right-inverse are more complicated since a notion of rank does not exist over rings.

Matrix inversion is the process of finding the matrix B that satisfies the prior equation for a given invertible matrix A.

Properties

The Invertible Matrix Theorem

Let A be a square n by n matrix over a field K (for example the field R of real numbers). The following statements are equivalent:

A is invertible, i.e. A has an inverse or is nonsingular.

A is row-equivalent to the n-by-n identity matrix I_n.

A is column-equivalent to the n-by-$n$identity matrix I_n.

A has n pivot positions.

$\det A \neq 0$. In general, a square matrix over a commutative ring is invertible if and only if its determinant is a unit in that ring.

A has full rank; that is, rank $A = n$.

The equation $Ax = 0$ has only the trivial solution $x = 0$

NullA = {0}

The equation $Ax = b$ has exactly one solution for each b in K^n.

The columns of A are linearly independent.

The columns of Aspan K^n

Col $A = K^n$

The columns of A form a basis of K^n.

The linear transformation mapping x to Ax is a bijection from K^n to K^n.

There is an n by n matrix B such that $AB = I_n = BA$.

The transpose A^T is an invertible matrix (hence rows of A are linearly independent, span K^n, and form a basis of K^n).

The number 0 is not an eigenvalue of A.

The matrix A can be expressed as a finite product of elementary matrices.

The matrix A has a left inverse (i.e. there exists a B such that BA = I) *or* a right inverse (i.e. there exists a C such that $AC = I$), in which case both left and right inverses exist and $B = C = A^{-1}$.

Other Properties

Furthermore, the following properties hold for an invertible matrix A:

- $(A^{-1})^{-1} = A$;
- $(kA)^{-1} = k^{-1}A^{-1}$ for nonzero scalar k;
- $(A^T)^{-1} = (A^{-1})^T$;
- For any invertible n-by-n matrices A and B, $(AB)^{-1} = B^{-1}A^{-1}$. More generally, if $A_1,...,A_k$ are invertible n-by-n matrices, then $(A_1A_2...A_{k-1}A_k)^{-1} = A_k^{-1}A_{k-1}^{-1}...A_2^{-1}A_1^{-1}$;
- $\det(A^{-1}) = \det(A)^{-1}$.

A matrix that is its own inverse, i.e. $A = A^{-1}$ and $A^2 = I$, is called an involution.

Density

Over the field of real numbers, the set of singular n-by-n matrices, considered as a subset of $R^{n\times n}$, is a null set, i.e., has Lebesguemeasure zero. This is true because singular matrices are the roots of the polynomial function in the entries of the matrix given by the determinant. Thus in the language of measure theory, almost all$n$-by-n matrices are invertible.

Furthermore the n-by-n invertible matrices are a denseopen set in the topological space of all n-by-n matrices. Equivalently, the set of singular matrices is closed and nowhere dense in the space of n-by-n matrices.

In practice however, one may encounter non-invertible matrices. And in numerical calculations, matrices which are invertible, but close to a non-invertible matrix, can still be problematic; such matrices are said to be ill-conditioned.

Chapter

2

Overview of Finite Mathematics

What is Finite Mathematics?

A very common question math faculty get is about the content of Math 124 Finite Math. Unfortunately, no simple answer can be given in twenty-five words or less. But that's why you are reading this, yes?

"Finite Math" is a catch-all title for a collection of topics that are anything but calculus. The purpose of the course is to give a survey of mathematical analysis techniques used in the working world, but you might also say that this course gives valuable experience at organizing information and then analyzing it. In a larger sense, it's also another way we use math to give people experience at analytical thinking. Business, accounting and computer majors tend to take this course, or are required to by their programme. It is an excellent choice for many education majors to fill their math requirements, but these people should check with their transfer institution first.

Real Number

In mathematics, a real number is a value that represents a quantity along a continuous line. The real numbers include all the rational numbers, such as the integer "5 and the fraction 4/3, and all the irrational numbers such as "2 (1.41421356... the square root of two, an irrational algebraic number) and π (3.14159265..., a transcendental number). Real numbers can be thought of as points on an infinitely long line called the number line or real line, where the points corresponding to integers are equally spaced. Any real number can be determined by a possibly infinite decimal representation such as that of 8.632, where each consecutive digit is measured in units one tenth the size of the previous one. The real line can be thought of as a part of the complex plane, and correspondingly, complex numbers include real numbers as a special case.

These descriptions of the real numbers are not sufficiently rigorous by the modern standards of pure mathematics. The discovery of a suitably rigorous definition of the real numbers — indeed, the realization that a better definition was needed — was one of the most important developments of 19th century mathematics. The currently standard axiomatic definition is that real numbers form the unique Archimedean complete totally ordered field $(R,+,\cdot,<)$, up to isomorphism, whereas popular constructive definitions of real numbers include declaring them as equivalence classes of Cauchy sequences of rational numbers, Dedekind cuts, or certain infinite "decimal representations", together with precise interpretations for the arithmetic operations and the order relation. These definitions are equivalent in the realm of classical mathematics.

The reals are uncountable, that is, while both the set of all natural numbers and the set of all real numbers are infinite sets, there can be no one-to-one function from the real numbers to the natural numbers: the cardinality of the set of all real numbers (denoted $\mathfrak{c}$ and called cardinality of the continuum) is strictly greater than the cardinality of the set of all natural numbers (denoted $\aleph_0$). The statement that there is no subset of the reals with cardinality strictly greater than $\aleph_0$ and strictly smaller than $\mathfrak{c}$ is known as the continuum hypothesis. It is known to be neither provable nor refutable using the axioms of Zermelo–Fraenkel set theory, the standard foundation of modern mathematics, provided ZF set theory is consistent.

Basic Properties

A real number may be either rational or irrational; either algebraic or transcendental; and either positive, negative, or zero. Real numbers are used to measure continuous quantities. They may be expressed by decimal representations that have an infinite sequence of digits to the right of the decimal point; these are often represented in the same form as 324.823122147... The ellipsis (three dots) indicate that there would still be more digits to come.

More formally, real numbers have the two basic properties of being an ordered field, and having the least upper bound property. The first says that real numbers comprise a field, with addition and multiplication as well as division by nonzero numbers, which can be totally ordered on a number line in a way compatible with addition and multiplication. The second says that if a nonempty set of real numbers has an upper bound, then it has a real least upper bound. The second condition distinguishes the real numbers from the rational numbers: for example, the set of rational numbers whose square is less than 2 is a set with an upper bound (e.g. 1.5) but no (rational) least upper bound: hence the rational numbers do not satisfy the least upper bound property.

In Physics

In the physical sciences, most physical constants such as the universal gravitational constant, and physical variables, such as position, mass, speed, and electric charge, are modelled using real numbers. In fact, the fundamental physical theories such as classical mechanics, electromagnetism, quantum mechanics, general relativity and the standard model are described using mathematical structures, typically smooth manifolds or Hilbert spaces, that are based on the real numbers although actual measurements of physical quantities are of finite accuracy and precision.

In some recent developments of theoretical physics stemming from the holographic principle, the Universe is seen fundamentally as an information store, essentially zeroes and ones, organized in much less geometrical fashion and manifesting itself as space-time and particle fields only on a more superficial level. This approach removes the real number system from its foundational role in physics and even prohibits the existence of infinite precision real numbers in the physical universe by considerations based on the Bekenstein bound.

In Computation

Computer arithmetic cannot directly operate on real numbers, but only on a finite subset of rational numbers, limited by the number of bits used to store them, whether as floating-point numbers or arbitrary precision numbers. However, computer algebra systems can operate on irrational quantities exactly by manipulating formulas for them (such as $\sqrt{2}$, $\arcsin\left(\frac{2}{23}\right)$, or $\int_0^1 x^x \, dx$) rather than their rational or decimal approximation; however, it is not in general possible to determine whether two such expressions are equal (the constant problem).

A real number is called *computable* if there exists an algorithm that yields its digits. Because there are only countably many algorithms, but an uncountable number of reals, almost all real numbers fail to be computable. Moreover, the equality of two computable numbers is an undecidable problem. Some constructivists accept the existence of only those reals that are computable. The set of definable numbers is broader, but still only countable.

Notation

Mathematicians use the symbol R (or alternatively, $\mathbb{R}$, the letter "*R*" in blackboard bold, Unicode $\mathbb{R}$ – *U*+211D) to represent the set of all real numbers (as this set is naturally endowed with a structure of field, the expression *field of the real numbers* is more frequently used than *set of all real numbers*). The notation R^n refers to the Cartesian product of n copies of R, which is an n-dimensional vector space over the field of the real numbers; this vector space may be identified to the n-dimensional space of Euclidean geometry as soon as a coordinate system has been chosen in the latter. For example, a value from R^3 consists of three real numbers and specifies the coordinates of a point in 3-dimensional space.

In mathematics, *real* is used as an adjective, meaning that the underlying field is the field of the real numbers (or *the real field*). For example *real matrix*, *real polynomial* and *real Lie algebra*. As a substantive, the term is used almost strictly in reference to the real numbers themselves (e.g., The "set of all reals").

History

Simple fractions have been used by the Egyptians around 1000 BC; the Vedic "Sulba Sutras" ("The rules of chords") in, ca. 600 BC, include what may be the first 'use' of irrational numbers. The concept of irrationality was implicitly accepted by early Indian mathematicians since Manava (c. 750–690 BC), who were aware that the square roots of certain numbers such as 2 and 61 could not be exactly determined. Around 500 BC, the Greek mathematicians led by Pythagoras realized the need for irrational numbers, in particular the irrationality of the square root of 2.

The Middle Ages brought the acceptance of zero, negative, integral, and fractional numbers, first by Indian and Chinese mathematicians, and then by Arabic mathematicians, who were also the first to treat irrational numbers as algebraic objects, which was made possible by the development of algebra. Arabic mathematicians merged the concepts of "number" and "magnitude" into a more general idea of real numbers. The Egyptian mathematician Abû Kâmil Shujâ ibn Aslam (c. 850–930) was the first to accept irrational numbers as solutions to quadratic equations or as coefficients in an equation, often in the form of square roots, cube roots and fourth roots.

In the 16th century, Simon Stevin created the basis for modern decimal notation, and insisted that there is no difference between rational and irrational numbers in this regard.

In the 17th century, Descartes introduced the term "real" to describe roots of a polynomial, distinguishing them from "imaginary" ones.

In the 18th and 19th centuries there was much work on irrational and transcendental numbers. Johann Heinrich Lambert (1761) gave the first flawed proof that π cannot be rational; Adrien-Marie Legendre (1794) completed the proof, and showed that π is not the square root of a rational number. Paolo Ruffini (1799) and Niels Henrik Abel (1842) both constructed proofs of Abel–Ruffini theorem: that the general quintic or higher equations cannot be solved by a general formula involving only arithmetical operations and roots.

Évariste Galois (1832) developed techniques for determining whether a given equation could be solved by radicals, which gave rise to the field of Galois theory. Joseph Liouville (1840) showed that neither e nor e^2 can be a root of an integer quadratic equation, and then established existence of transcendental numbers, the proof being subsequently displaced by Georg Cantor (1873). Charles Hermite (1873) first proved that e is transcendental, and Ferdinand von Lindemann (1882), showed that π is transcendental. Lindemann's proof was much simplified by Weierstrass (1885), still further by David Hilbert (1893), and has finally been made elementary by Adolf Hurwitz and Paul Gordan.

The development of calculus in the 18th century used the entire set of real numbers without having defined them cleanly. The first rigorous definition was given by Georg Cantor in 1871. In 1874 he showed that the set of all real numbers is uncountably infinite but the set of all algebraic numbers is countably infinite. Contrary to widely held beliefs, his first method was not his famous diagonal argument, which he published in 1891.

Construction of the real numbers

In mathematics, there are several ways of defining the real number system as an ordered field. The *synthetic* approach gives a list of axioms for the real numbers as a *complete ordered field*. Under the usual axioms of set theory, one can show that these axioms are categorical, in the sense that there is a model for the axioms, and any two such models are isomorphic. Any one of these models must be explicitly constructed, and most of these models are built using the basic properties of the rational number system as an ordered field.

Synthetic Approach

The synthetic approach axiomatically defines the real number system as a complete ordered field. Precisely, this means the following. A *model for the real number system* consists of a set R, two distinct elements 0 and 1 of R, two binary operations + and * on R (called *addition* and *multiplication*, resp.), a binary relation $\leq$ on R, satisfying the following properties.

1. $(R, +, *)$ forms a field. In other words,
   — For all x, y, and z in R, $x + (y + z) = (x + y) + z$ and $x * (y * z) = (x * y) * z$. (associativity of addition and multiplication)
   — For all x and y in R, $x + y = y + x$ and $x * y = y * x$. (commutativity of addition and multiplication)
   — For all x, y, and z in R, $x * (y + z) = (x * y) + (x * z)$. (distributivity of multiplication over addition)
   — For all x in R, $x + 0 = x$. (existence of additive identity)
   — 0 is not equal to 1, and for all x in R, $x * 1 = x$. (existence of multiplicative identity)
   — For every x in R, there exists an element “x in R, such that x + (“x) = 0. (existence of additive inverses)
   — For every $x \neq 0$ in R, there exists an element x^{-1} in R, such that $x * x^{-1} = 1$. (existence of multiplicative inverses)
2. (R, $\leq$) forms a totally ordered set. In other words,
   — For all x in R, $x \leq x$. (reflexivity)
   — For all x and y in R, if $x \leq y$ and $y \leq x$, then $x = y$. (antisymmetry)
   — For all x, y, and z in R, if $x \leq y$ and $y \leq z$, then $x \leq z$. (transitivity)
   — For all x and y in R, $x \leq y$ or $y \leq x$. (totalness)

3. The field operations + and * on R are compatible with the order $\leq$. In other words,
   — For all x, y and z in R, if $x \leq y$, then $x + z \leq y + z$. (preservation of order under addition)
   — For all x and y in R, if $0 \leq x$ and $0 \leq y$, then $0 \leq x * y$ (preservation of order under multiplication)
4. The order $\leq$ is *complete* in the following sense: every non-empty subset of R bounded above has a least upper bound. In other words,
   — If A is a non-empty subset of R, and if A has an upper bound, then A has a least upper bound u, such that for every upper bound v of A, $u \leq v$.

The final axiom, defining the order as Dedekind-complete, is most crucial. Without this axiom, we simply have the axioms which define a totally ordered field, and there are many non-isomorphic models which satisfy these axioms. This axiom implies that the Archimedean property applies for this field. Therefore, when the completeness axiom is added, it can be proved that any two models must be isomorphic, and so in this sense, there is only one complete ordered Archimedean field.

When we say that any two models of the above axioms are isomorphic, we mean that for any two models $(R, 0_R, 1_R, +_R, *_R, \leq_R)$ and $(S, 0_S, 1_S, +_S, *_S, \leq_S)$, there is a bijection $f : R \to S$ preserving both the field operations and the order. Explicitly,

- f is both injective and surjective.
- $f(0_R) = 0_S$ and $f(1_R) = 1_S$.
- For all x and y in R, $f(x +_R y) = f(x) +_S f(y)$ and $f(x *_R y) = f(x) *_S f(y)$.
- For all x and y in R, $x \leq_R y$ if and only if $f(x) \leq_S f(y)$.

Explicit Constructions of Models

We shall not prove that any models of the axioms are isomorphic. Such a proof can be found in any number of modern analysis or set theory textbooks. We will sketch the basic definitions and properties of a number of constructions, however, because each of these is important for both mathematical and historical reasons. The first three, due to Georg Cantor/Charles Méray, Richard Dedekind and Karl Weierstrass/Otto Stolz all occurred within a few years of each other. Each has advantages and disadvantages. A major motivation in all three cases was the instruction of mathematics students.

Construction from Cauchy Sequences

If we have a space where Cauchy sequences are meaningful (such as a 'rational' metric space, i.e., a space in which distance is defined and takes rational values, or more generally a uniform space), a standard procedure to force all Cauchy sequences to converge is adding new points to the space (a process called completion). By starting with rational numbers and the metric $d(x,y) = |x$ " $y|$, we can construct the real numbers, as will be detailed below. (A different metric on the rationals could result in the *p*-adic numbers instead.)

Let R be the set of Cauchy sequences of rational numbers. That is, sequences

$x_1,x_2,x_3,\ldots$ of rational numbers such that for every rational $\varepsilon > 0$, there exists an integer N such that for all natural numbers $m,n>N$, $|x_m-x_n|<\varepsilon$. Here the vertical bars denote the absolute value. Cauchy sequences (x) and (y) can be added and multiplied as follows:

$$(x_n) + (y_n) = (x_n + y_n)$$

$$(x_n) \times (y_n) = (x_n \times y_n).$$

Two Cauchy sequences are called *equivalent* if and only if the difference between them tends to zero. Comparison between two cauchy sequences is possible as such : $(x_n) \geq (y_n)$ if and only if x is equivalent to y or there exists an integer N such that $x_n \geq y_n$ for all $n>N$.

This does indeed define an equivalence relation, it is compatible with the operations defined above, and the set R of all equivalence classes can be shown to satisfy all the usual axioms of the real numbers. This is remarkable because not all of these axioms necessarily apply to the rational numbers, which are being used to construct the sequences themselves. We can embed the rational numbers into the reals by identifying the rational number r with the equivalence class of the sequence $(r,r,r,\ldots)$.

The only real number axiom that does not follow easily from the definitions is the completeness of ≤, i.e. the least upper bound property. It can be proved as follows: Let S be a non-empty subset of R and U be an upper bound for S. Substituting a larger value if necessary, we may assume U is rational. Since S is non-empty, there is a rational number L such that $L<s$ for some s in S. Now define sequences of rationals (u_n) and (l_n) as follows:

Set $u_0 = U$ and $l_0 = L$.

For each n consider the number:

$$m_n = (u_n + l_n)/2$$

If m_n is an upper bound for S set:

$$u_{n+1} = m_n \text{ and } l_{n+1} = l_n$$

Otherwise set:

$$l_{n+1} = m_n \text{ and } u_{n+1} = u_n$$

This obviously defines two Cauchy sequences of rationals, and so we have real numbers $l = (l_n)$ and $u = (u_n)$. It is easy to prove, by induction on n that:

u_n is an upper bound for S for all n

and:

l_n is never an upper bound for S for any n

Thus u is an upper bound for S. To see that it is a least upper bound, notice that the limit of $(u_n$ " $l_n)$ is 0, and so $l = u$. Now suppose $b<u = l$ is a smaller upper bound for S. Since (l_n) is monotonic increasing it is easy to see that $b<l_n$ for some n. But l_n is not an upper bound for S and so neither is b. Hence u is a least upper bound for S and $\leq$ is complete.

A practical and concrete representative for an equivalence class representing a real number is provided by the representation to base b – in practice, b is usually 2 (binary), 8 (octal), 10 (decimal) or 16 (hexadecimal). For example, the number $\pi = 3.14159...$ corresponds to the Cauchy sequence (3,3.1,3.14,3.141,3.1415,...). Notice that the sequence (0,0.9,0.99,0.999,0.9999,...) is equivalent to the sequence (1,1.0,1.00,1.000,1.0000,...); this shows that 0.999... = 1.

An advantage of this approach is that it does not use the linear order of the rationals, only the metric. Hence it generalizes to other metric spaces.

Construction by Dedekind Cuts

A Dedekind cut in an ordered field is a partition of it, (A, B), such that A is nonempty and closed downwards, B is nonempty and closed upwards, and A contains no greatest element. Real numbers can be constructed as Dedekind cuts of rational numbers.

For convenience we may take the lower set A as the representative of any given Dedekind cut (A,B), since A completely determines B. By doing this we may think intuitively of a real number as being represented by the set of all smaller rational numbers. In more detail, a real number is any subset of the set Q of rational numbers that fulfills the following conditions:

1. r is not empty
2. $r \neq Q$
3. r is closed downwards. In other words, for all $x, y \in Q$ such that $x < y$, if $y \in r$ then $x \in r$
4. r contains no greatest element. In other words, there is no such that for all $y \leq x$,
   - We form the set R of real numbers as the set A of Q all Dedekind cuts of , and define a total ordering on the real numbers as follows: $x \leq y \Leftrightarrow x \subseteq y$
   - We embed the rational numbers into the reals by identifying the rational number q with the set of all smaller rational numbers $\{x \in \mathrm{Q} : x < q\}$. Since the rational numbers are dense, such a set can have no greatest element and thus fulfills the conditions for being a real number laid out above.
   - Addition. $A + B := \{a + b : a \in A \wedge b \in B\}$
   - Subtraction. $A - B := \{a - b : a \in A \wedge b \in (\mathrm{Q} \setminus B)\}$ where $\mathrm{Q} \setminus B$ denotes the relative complement of B in Q, $\{x : x \in \mathrm{Q} \wedge x \notin B\}$
   - Negation is a special case of subtraction:

     $-B := \{a - b : a < 0 \wedge b \in (Q \setminus B)\}$
   - Defining multiplication is less straightforward.
     - if $A, B \geq 0$ then

       $A \times B := \{a \times b : a \geq 0 \wedge a \in A \wedge b \geq 0 \wedge b \in B\} \cup \{x \in Q : x < 0\}$
     - if either A or is negative, we use the identities $A \times B = -(A \times -B) = -(-A \times B) = (-A \times -B)$ to convert A and/or B to positive numbers and then apply the definition above.
     - We define division in a similar manner:
     - if $A \geq 0$ and $B > 0$ then

       $A / B := \{a / b : a \in A \wedge b \in (\mathrm{Q} \setminus B)\}$
     - if either A or B is negative, we use the identities

       $A / B = -(A / -B) = -(-A / B) = -A / -B$

       to convert A to a non-negative number and/or B to a positive number and then apply the definition above.

- Supremum. If a nonempty set S of real numbers has any upper bound in R, then it has a least upper bound in R that is equal to $\bigcup S$.

As an example of a Dedekind cut representing an irrational number, we may take the positive square root of 2. This can be defined by the set $A = \{x \in \mathbb{Q} : x < 0 \vee x \times x < 2\}$. It can be seen from the definitions above that A is a real number, and that $A \times A = 2$. However, neither claim is immediate. Showing that A is real requires showing that for any positive rational x with $x \times x < 2$, there is a rational y with $x < y$ and $y \times y < 2$. The choice $y = \frac{2x+2}{x+2}$ works. Then $A \times A \leq 2$ but to show equality requires showing that if r is any rational number less than 2, then there is positive x in A with $r < x \times x$.

An advantage of this construction is that each real number corresponds to a unique cut.

Stevin's Construction

It has been known since Simon Stevin that real numbers can be represented by decimals. We can take the infinite decimal expansion to be the definition of a real number, defining expansions like 0.9999... and 1.0000... to be equivalent, and define the arithmetical operations formally. This is equivalent to the constructions by Cauchy sequences or Dedekind cuts and incorporates an explicit modulus of convergence. Similarly, another radix can be used. Weierstrass attempted to construct the reals but did not entirely succeed. He pointed out that they need only be thought of as complete aggregates (sets) of units and unit fractions.

This construction has the advantage that it is close to the way we are used to thinking of real numbers and suggests series expansions for functions. A standard approach to show that all models of a complete ordered field are isomorphic is to show that any model is isomorphic to this one because we can systematically build a decimal expansion for each element.

Construction Using Hyperreal Numbers

As in the hyperreal numbers, one constructs the hyperrationals $^*Q$ from the rational numbers by means of an ultrafilter. Here a hyperrational is by definition a ratio of two hyperintegers. Consider the ring B of all limited (i.e. finite) elements in $^*Q$. Then B has a unique maximal ideal I, the infinitesimal numbers. The quotient ring

B/I gives the field R of real numbers. Note that B is not an internal set in $^*Q$. Note that this construction uses a non-principal ultrafilter over the set of natural numbers, the existence of which is guaranteed by the axiom of choice.

It turns out that the maximal ideal respects the order on $^*Q$. Hence the resulting field is an ordered field. Completeness can be proved in a similar way to the construction from the Cauchy sequences.

Construction from Surreal Numbers

Every ordered field can be embedded in the surreal numbers. The real numbers form a maximal subfield that is Archimedean (meaning that no real number is infinitely large). This embedding is not unique, though it can be chosen in a canonical way.

Construction from Z (Eudoxus Reals)

A relatively less known construction allows to define real numbers using only the additive group of integers $\mathbb{Z}$ with different versions. The construction has been formally verified by the Isar Math Lib project. Shenitzer and Arthan refer to this construction as the *Eudoxus reals*.

Let an almost homomorphism be a map $f:\mathbb{Z}\to\mathbb{Z}$ such that the set $\{f(n+m)-f(m)-f(n): n,m\in\mathbb{Z}\}$ is finite. We say that two almost homomorphisms f,g are almost equal if the set $\{f(n)-g(n): n\in\mathbb{Z}\}$ is finite. This defines an equivalence relation on the set of almost homomorphisms. Real numbers are defined as the equivalence classes of this relation.

To add real numbers defined this way we add the almost homomorphisms that represent them. Multiplication of real numbers corresponds to composition of almost homomorphisms. If $[f]$ denotes the real number represented by an almost homomorphism f we say that $0\le[f]$ if f is bounded or f takes an infinite number of positive values on $\mathbb{Z}^+$. This defines the linear order relation on the set of real numbers constructed this way.

Other Constructions

Few mathematical structures have undergone as many revisions or have been presented in as many guises as the real numbers. Every generation reexamines the reals in the light of its values and mathematical objectives. A number of constructions have been given. As a reviewer of one noted: "The details are all included, but as usual they are tedious and not too instructive."

Properties

Completeness

A main reason for using real numbers is that the reals contain all limits. More precisely, every sequence of real numbers having the property that consecutive terms of the sequence become arbitrarily close to each other necessarily has the property that after some term in the sequence the remaining terms are arbitrarily close to some specific real number. In mathematical terminology, this means that the reals are complete (in the sense of metric spaces or uniform spaces, which is a different sense than the Dedekind completeness of the order in the previous section). This is formally defined in the following way:

A sequence (x_n) of real numbers is called a *Cauchy sequence* if for any $\varepsilon > 0$ there exists an integer N (possibly depending on ε) such that the distance $|x_n - x_m|$ is less than ε for all n and m that are both greater than N. In other words, a sequence is a Cauchy sequence if its elements x_n eventually come and remain arbitrarily close to each other.

A sequence (x_n) *converges to the limit* x if for any $\varepsilon > 0$ there exists an integer N (possibly depending on ε) such that the distance $|x_n - x|$ is less than ε provided that n is greater than N. In other words, a sequence has limit x if its elements eventually come and remain arbitrarily close to x. Notice that every convergent sequence is a Cauchy sequence. The converse is also true: Every Cauchy sequence of real numbers is convergent to a real number.

That is, the reals are complete. Note that the rationals are not complete. For example, the sequence (1, 1.4, 1.41, 1.414, 1.4142, 1.41421, ...), where each term adds a digit of the decimal expansion of the positive square root of 2, is Cauchy but it does not converge to a rational number. (In the real numbers, in contrast, it converges to the positive square root of 2.)

The existence of limits of Cauchy sequences is what makes calculus work and is of great practical use. The standard numerical test to determine if a sequence has a limit is to test if it is a Cauchy sequence, as the limit is typically not known in advance.

For example, the standard series of the exponential function

$$e^x = \sum_{n=0}^{\infty} \frac{x^n}{n!}$$

converges to a real number because for every x the sums

$$\sum_{n=N}^{M} \frac{x^n}{n!}$$

can be made arbitrarily small by choosing N sufficiently large. This proves that the sequence is Cauchy, so we know that the sequence converges even if the limit is not known in advance.

"The Complete Ordered Field"

The real numbers are often described as "the complete ordered field", a phrase that can be interpreted in several ways.

First, an order can be lattice-complete. It is easy to see that no ordered field can be lattice-complete, because it can have no largest element (given any element z, $z + 1$ is larger), so this is not the sense that is meant.

Additionally, an order can be Dedekind-complete, as defined in the section Axioms. The uniqueness result at the end of that section justifies using the word "the" in the phrase "complete ordered field" when this is the sense of "complete" that is meant. This sense of completeness is most closely related to the construction of the reals from Dedekind cuts, since that construction starts from an ordered field (the rationals) and then forms the Dedekind-completion of it in a standard way.

These two notions of completeness ignore the field structure. However, an ordered group (in this case, the additive group of the field) defines a uniform structure, and uniform structures have a notion of completeness (topology); the description in the section Completeness above is a special case. (We refer to the notion of completeness in uniform spaces rather than the related and better known notion for metric spaces, since the definition of metric space relies on already having a characterisation of the real numbers.) It is not true that R is the *only* uniformly complete ordered field, but it is the only uniformly complete *Archimedean field*, and indeed one often hears the phrase "complete Archimedean field" instead of "complete ordered field". Every uniformly complete Archimedean field must also be Dedekind-complete (and vice versa, of course), justifying using "the" in the phrase "the complete Archimedean field". This sense of completeness is most closely related to the construction of the reals from Cauchy sequences (the construction carried out in full in this article), since it starts with an Archimedean field (the rationals) and forms the uniform completion of it in a standard way.

But the original use of the phrase "complete Archimedean field" was by David Hilbert, who meant still something else by it. He meant that the real numbers form the *largest* Archimedean field in the sense that every other Archimedean field is a subfield of R. Thus R is "complete" in the sense that nothing further can be added to it without making it no longer an Archimedean field. This sense of completeness is most closely related to the construction of the reals from surreal numbers, since that construction starts with a proper class that contains every ordered field (the surreals) and then selects from it the largest Archimedean subfield.

Advanced Properties

The reals are uncountable, that is, there are strictly more real numbers than natural numbers, even though both sets are infinite. In fact, the cardinality of the reals equals that of the set of subsets (i.e., the power set) of the natural numbers, and Cantor's diagonal argument states that the latter set's cardinality is strictly bigger than the cardinality of N. Since only a countable set of real numbers can be algebraic, almost all real numbers are transcendental. The non-existence of a subset of the reals with cardinality strictly between that of the integers and the reals is known as the continuum hypothesis. The continuum hypothesis can neither be proved nor be disproved; it is independent from the axioms of set theory.

As a topological space, the real numbers are separable. This is because the set of rationals, which is countable, is dense in the real numbers. The irrational numbers are also dense in the real numbers, however they are uncountable and have the same cardinality as the reals. The real numbers form a metric space: the distance between x and y is defined as the absolute value $|x " y|$. By virtue of being a totally ordered set, they also carry an order topology; the topology arising from the metric and the one arising from the order are identical, but yield different presentations for the topology – in the order topology as intervals, in the metric topology as epsilon-balls. The Dedekind cuts construction uses the order topology presentation, while the Cauchy sequences construction uses the metric topology presentation. The reals are a contractible (hence connected and simply connected), separable and complete metric space of Hausdorff dimension 1. The real numbers are locally compact but not compact. There are various properties that uniquely specify them; for instance, all unbounded, connected, and separable order topologies are necessarily homeomorphic to the reals.

Every nonnegative real number has a square root in *R*, although no negative number does. This shows that the order on *R* is determined by its algebraic structure. Also, every polynomial of odd degree admits at least one real root: these two properties make *R* the premier example of a real closed field. Proving this is the first half of one proof of the fundamental theorem of algebra.

The reals carry a canonical measure, the Lebesgue measure, which is the Haar measure on their structure as a topological group normalised such that the unit interval [0,1] has measure 1. There exist sets of real numbers that are not Lebesgue measurable, e.g. Vitali sets.

The supremum axiom of the reals refers to subsets of the reals and is therefore a second-order logical statement. It is not possible to characterize the reals with first-order logic alone: the Löwenheim–Skolem theorem implies that there exists a countable dense subset of the real numbers satisfying exactly the same sentences in first order logic as the real numbers themselves.

The set of hyperreal numbers satisfies the same first order sentences as *R*. Ordered fields that satisfy the same first-order sentences as *R* are called nonstandard models of *R*. This is what makes nonstandard analysis work; by proving a first-order statement in some nonstandard model (which may be easier than proving it in *R*), we know that the same statement must also be true of *R*.

The field R of real numbers is an extension field of the field *Q* of rational numbers, and *R* can therefore be seen as a vector space over *Q*. Zermelo–Fraenkel set theory with the axiom of choice guarantees the existence of a basis of this vector space: there exists a set *B* of real numbers such that every real number can be written uniquely as a finite linear combination of elements of this set, using rational coefficients only, and such that no element of *B* is a rational linear combination of the others. However, this existence theorem is purely theoretical, as such a base has never been explicitly described.

The well-ordering theorem implies that the real numbers can be well-ordered if the axiom of choice is assumed: there exists a total order on R with the property that every non-empty subset of *R* has a least element in this ordering. (The standard ordering $\leq$ of the real numbers is not a well-ordering since e.g. an open interval does not contain a least element in this ordering.) Again, the existence of such a well-ordering is purely theoretical, as it cannot be explicitly described.

Generalizations and Extensions

The real numbers can be generalized and extended in several different directions:

- The complex numbers contain solutions to all polynomial equations and hence are an algebraically closed field unlike the real numbers. However, the complex numbers are not an ordered field.
- The affinely extended real number system adds two elements $+\infty$ and $-\infty$. It is a compact space. It is no longer a field, not even an additive group, but it still has a total order; moreover, it is a complete lattice.
- The real projective line adds only one value ∞. It is also a compact space. Again, it is no longer a field, not even an additive group. However, it allows division of a non-zero element by zero. It is not ordered any more.
- The long real line pastes together $\aleph_1{}^* + \aleph_1$ copies of the real line plus a single point (here $\aleph_1{}^*$ denotes the reversed ordering of $\aleph_1$) to create an ordered set that is "locally" identical to the real numbers, but somehow longer; for instance, there is an order-preserving embedding of $\aleph_1$ in the long real line but not in the real numbers. The long real line is the largest ordered set that is complete and locally Archimedean. As with the previous two examples, this set is no longer a field or additive group.
- Ordered fields extending the reals are the hyperreal numbers and the surreal numbers; both of them contain infinitesimal and infinitely large numbers and are therefore Non-Archimedean ordered fields.
- Self-adjoint operators on a Hilbert space (for example, self-adjoint square complex matrices) generalize the reals in many respects: they can be ordered (though not totally ordered), they are complete, all their eigenvalues are real and they form a real associative algebra. Positive-definite operators correspond to the positive reals and normal operators correspond to the complex numbers.

"Reals" in set Theory

In set theory, specifically descriptive set theory, the Baire space is used as a surrogate for the real numbers since the latter have some topological properties (connectedness) that are a technical inconvenience. Elements of Baire space are referred to as “reals”.

Real Numbers and Logic

The real numbers are most often formalized using the Zermelo–Fraenkel axiomatization of set theory, but some mathematicians study the real numbers with other logical foundations of mathematics. In particular, the real numbers are also studied in reverse mathematics and in constructive mathematics.

Abraham Robinson’s theory of nonstandard or hyperreal numbers extends the set of the real numbers by infinitesimal numbers, which allows building infinitesimal calculus in a way closer to the usual intuition of the notion of limit. Edward Nelson’s internal set theory is a non-Zermelo–Fraenkel set theory that considers non-standard real numbers as elements of the set of the reals (and not of an extension of it, as in Robinson’s theory).

The continuum hypothesis posits that the cardinality of the set of the real numbers is $\aleph_1$, i.e. the smallest infinite cardinal number after $\aleph_0$, the cardinality of the integers. Paul Cohen proved in 1963 that it is an axiom independent of the other axioms of set theory; that is, one may choose either the continuum hypothesis or its negation as an axiom of set theory, without contradiction.

Exponentiation

Exponentiation is a mathematical operation, written as b^n, involving two numbers, the base b and the exponent (or index or power) n. When n is a positive integer, exponentiation corresponds to repeated multiplication; in other words, a product of n factors, each of which is equal to b (the product itself can also be called power):

$$b^n = \underbrace{b \times \cdots \times b}_{n}$$

just as multiplication by a positive integer corresponds to repeated addition:

$$b \times n = \underbrace{b + \cdots + b}_{n}$$

The exponent is usually shown as a superscript to the right of the base. The exponentiation b^n can be read as: *b raised to the n-th power*, *b raised to the power of n*, or *b raised by the exponent of n*, most briefly as *b to the n*. Some exponents have their own pronunciation: for example, b^2 is usually read as *b squared* and b^3 as *b cubed.*

The power b^n can be defined also when n is a negative integer, for nonzero b. No natural extension to all real b and n exists, but when the base b is a positive real number, b^n can be defined for all real and even complex exponents n via the exponential function e^z. Trigonometric functions can be expressed in terms of complex exponentiation.

Exponentiation where the exponent is a matrix is used for solving systems of linear differential equations.

Exponentiation is used pervasively in many other fields, including economics, biology, chemistry, physics, and computer science, with applications such as compound interest, population growth, chemical reaction kinetics, wave behaviour, and public key cryptography.

Background and Terminology

The expression $b^2 = b \cdot b$ is called the square of b because the area of a square with side-length b is b^2.

The expression $b^3 = b \cdot b \cdot b$ is called the cube, because the volume of a cube with side-length b is b^3.

So 3^2 is pronounced "three squared", and 2^3 is "two cubed".

The exponent says how many copies of the base are multiplied together. For example, $3^5 = 3 \cdot 3 \cdot 3 \cdot 3 \cdot 3 = 243$. The base 3 appears 5 times in the repeated multiplication, because the exponent is 5. Here, 3 is the *base*, 5 is the *exponent*, and 243 is the *power* or, more specifically, *the fifth power of 3, 3 raised to the fifth power*, or *3 to the power of 5.*

The word "raised" is usually omitted, and very often "power" as well, so 3^5 is typically pronounced "three to the fifth" or "three to the five".

Exponentiation may be generalized from integer exponents to more general types of numbers.

Integer Exponents

The exponentiation operation with integer exponents requires only elementary algebra.

Positive Integer Exponents

Formally, powers with positive integer exponents may be defined by the initial condition

$$b^1 = b$$

and the recurrence relation

$$b^{n+1} = b^n \cdot b \cdot$$

From the associativity of multiplication, it follows that for any positive integers m and n,

$$b^{m+n} = b^m \cdot b^n$$

Arbitrary Integer Exponents

For non-zero b and positive n, the recurrence relation from the previous subsection can be rewritten as

$$b^n = \frac{b^{n+1}}{b}$$

By defining this relation as valid for all integer n and nonzero b, it follows that

$$b^0 = \frac{b^1}{b} = 1$$

$$b^{-1} = \frac{b^0}{b} = \frac{1}{b}$$

and more generally,

$$b^{-n} = \frac{1}{b^n}$$

for any nonzero b and any nonnegative integer n (and indeed any integer n).

The following observations may be made:

- Any number raised by the exponent 1 is the number itself.
- Any nonzero number raised by the exponent 0 is 1; one interpretation of these powers is as empty products.
- These equations do not decide the value of 0^0. This is discussed below.
- Raising 0 by a negative exponent would imply division by 0, so it is left undefined.

The identity

$$b^{m+n} = b^m b^n$$

initially defined only for positive integers m and n, holds for arbitrary integers m and n, with the constraint that m and n must both be positive when b is zero.

Identities and Properties

The following identities hold, provided that the base is non-zero whenever the integer exponent is not positive:

$$b^{m+n} = b^m \cdot b^n$$
$$(b^m)^n = b^{m \cdot n}$$
$$(b \cdot c)^n = b^n \cdot c^n$$

Exponentiation is not commutative. This contrasts with addition and multiplication, which are. For example, 2 + 3 = 3 + 2 = 5 and 2 · 3 = 3 · 2 = 6, but 2^3 = 8, whereas 3^2 = 9.

Exponentiation is not associative either. Addition and multiplication are. For example, (2 + 3) + 4 = 2 + (3 + 4) = 9 and (2 · 3) · 4 = 2 · (3 · 4) = 24, but 2^3 to the 4 is 8^4 or 4,096, whereas 2 to the 3^4 is 2^{81} or 2,417,851,639,229,258,349,412,352. Without parentheses to modify the order of calculation, by convention the order is top-down, not bottom-up:

$$b^{p^q} = b^{(p^q)} \neq (b^p)^q = b^{(p \cdot q)} = b^{p \cdot q}$$

(In the above formula, one could think that it is the (smaller) size of the last exponent q with respect to the p, which indicates the order of evaluation, but the convention also applies to the linearized writing *b^p^q = b^(p^q) ≠ (b^p)^q*, as it is used for example in computer algebra systems.)

Particular Bases

Powers of ten: In the base ten (decimal) number system, integer powers of 10 are written as the digit 1 followed or preceded by a number of zeroes determined by the sign and magnitude of the exponent. For example, 10^3 = 1,000 and 10^{-4} = 0.0001.

Exponentiation with base 10 is used in scientific notation to denote large or small numbers. For instance, 299,792,458 m/s (the speed of light in vacuum, in metre per second) can be written as 2.99792458×10^8 m/s and then approximated as 2.998×10^8 m/s. SI prefixes based on powers of 10 are also used to describe small or large quantities. For example, the prefix kilo means 10^3 = 1,000, so a kilometre is 1,000 metres.

Powers of Two

The positive powers of 2 are important in computer science because there are 2^n possible values for an n-bitbinary variable.

Powers of 2 are important in set theory since a set with n members has a power set, or set of all subsets of the original set, with 2^n members.

The negative powers of 2 are commonly used, and the first two have special names: half, and quarter.

In the base 2 (binary) number system, integer powers of 2 are written as 1 followed or preceded by a number of zeroes determined by the sign and magnitude of the exponent. For example, two to the power of three is written as 1000 in binary.

Powers of One

The integer powers of one are all one: $1^n = 1$.

Powers of Zero

If the exponent is positive, the power of zero is zero: $0^n = 0$, where $n > 0$.

If the exponent is negative, the power of zero (0^n, where $n < 0$) is undefined, because division by zero is implied.

If the exponent is zero, some authors define $0^0 = 1$, whereas others leave it undefined, as discussed below.

Powers of Minus One

If n is an even integer, then $(-1)^n = 1$.

If n is an odd integer, then $(-1)^n = -1$.

Because of this, powers of -1 are useful for expressing alternating sequences.

Large Exponents

The limit of a sequence of powers of a number greater than one diverges, in other words they grow without bound:

$$b^n \to \infty \text{ as } n \to \infty \text{ when } b > 1$$

This can be read as "b to the power of n tends to $+\infty$ as n tends to infinity when b is greater than one".

Powers of a number with absolute value less than one tend to zero:

$$b^n \to 0 \text{ as } n \to \infty \text{ when } |b| < 1$$

Any power of one is always itself:

$$b^n = 1 \text{ for all } n \text{ if } b = 1$$

If the number b varies tending to 1 as the exponent tends to infinity then the limit is not necessarily one of those above. A particularly important case is

$$(1 + 1/n)^n \to e \text{ as } n \to \infty$$

Other limits, in particular of those tending to indeterminate forms, are described in limits of powers below.

Rational Exponents

An n-th root of a number b is a number x such that $x^n = b$.

If b is a positive real number and n is a positive integer, then there is exactly one positive real solution to $x^n = b$. This solution is called the principal n-th root of b. It is denoted $^n\sqrt{b}$, where $\sqrt{}$ is the radical symbol; alternatively, it may be written $b^{1/n}$. For example: $4^{1/2} = 2$, $8^{1/3} = 2$,

When one speaks of *then*-th root of a positive real number$b$, one usually means the principal n-th root.

If n is even, then $x^n = b$ has two real solutions if b is positive, which are the positive and negative nth roots. The equation has no solution in real numbers if b is negative.

If n is odd, then $x^n = b$ has one real solution. The solution is positive if b is positive and negative if b is negative.

Rational powers m/n, where m/n is in lowest terms, are positive if m is even, negative for negative b if m and n are odd, and can be either sign if b is positive and n is even. $(-27)^{1/3} = -3$, $(-27)^{2/3} = 9$, and $4^{3/2}$ has two roots 8 and –8. Since there is no real number x such that $x^2 = -1$, the definition of $b^{m/n}$ when b is negative and n is even must use the imaginary unit i, as described more fully in the section Powers of complex numbers.

A power of a positive real number b with a rational exponent m/n in lowest terms satisfies

$$b^{\frac{m}{n}} = \left(b^m\right)^{\frac{1}{n}} = \sqrt[n]{b^m}$$

where m is an integer and n is a positive integer.

Care needs to be taken when applying the power law identities with negative nth roots. For instance, $-27 = (-27)^{((2/3)(3/2))} = ((-27)^{2/3})^{3/2} = 9^{3/2} = 27$ is clearly wrong. The problem here occurs in taking the positive square root rather than the negative one at the last step, but in general the same sorts of problems occur as described for complex numbers in the section Failure of power and logarithm identities.

Real Exponents

The identities and properties shown above for integer exponents are true for positive real numbers with non-integer exponents as well.

However the identity

$$(b^r)^s = b^{r \cdot s}$$

cannot be extended consistently to where b is a negative real number, see Real exponents with negative bases. The failure of this identity is the basis for the problems with complex number powers detailed under failure of power and logarithm identities.

The extension of exponentiation to real powers of positive real numbers can be done either by extending the rational powers to reals by continuity, or more usually by using the exponential function and its inverse, the natural logarithm.

Limits of Rational Exponents

Since any irrational number can be approximated by a rational number, exponentiation of a positive real number b with an arbitrary real exponent x can be defined by continuity with the rule·

$$b^x = \lim_{r \to x} b^r \quad (r \in \mathbb{Q}, x \in \mathbb{R})$$

where the limit as r gets close to x is taken only over rational values of r. This limit only exists for positive b. The (ε, ξ)-definition of limit is used, this involves showing that for any desired accuracy of the result b^x one can choose a sufficiently small interval around x so all the rational powers in the interval are within the desired accuracy.

For example, if $x = \pi$, the nonterminating decimal representation $\pi = 3.14159\ldots$ can be used (based on strict monotonicity of the rational power) to obtain the intervals bounded by rational powers

$[b^3, b^4]$, $[b^{3.1}, b^{3.2}]$, $[b^{3.14}, b^{3.15}]$, $[b^{3.141}, b^{3.142}]$, $[b^{3.1415}, b^{3.1416}]$, $[b^{3.14159}, b^{3.14160}]$, ...

The bounded intervals converge to a unique real number, denoted by b^π. This technique can be used to obtain any irrational power of b. The function $f(x) = b^x$ is thus defined for any real number x.

The exponential function

The important mathematical constant e, sometimes called Euler's number, is approximately equal to 2.718 and is the base of the natural logarithm. Although exponentiation of e could, in principle, be treated the same as exponentiation of any other real number, such exponentials turn out to have particularly elegant and useful properties. Among other things, these properties allow exponentials of e to be generalized in a natural way to other types of exponents, such as complex numbers or even matrices, while coinciding with the familiar meaning of exponentiation with rational exponents.

As a consequence, the notation e^x usually denotes a generalized exponentiation definition called the exponential function, $\exp(x)$, which can be defined in many equivalent ways, for example by:

$$\exp(x) = \lim_{n\to\infty}\left(1+\frac{x}{n}\right)^n$$

Among other properties, exp satisfies the exponential identity:

$$\exp(x+y) = \exp(x)\cdot\exp(y)$$

The exponential function is defined for all integer, fractional, real, and complex values of x. It can even be used to extend exponentiation to some nonnumerical entities such as square matrices (in which case the exponential identity only holds when x and y commute).

Since $\exp(1)$ is equal to e and $\exp(x)$ satisfies the exponential identity, it immediately follows that $\exp(x)$ coincides with the repeated-multiplication definition of e^x for integer x, and it also follows that rational powers denote (positive) roots as usual, so exp(x) coincides with the e^x definitions in the previous section for all real x by continuity.

Powers Via Logarithms

The natural logarithm $\ln(x)$ is the inverse of the exponential function e^x. It is defined for $b > 0$, and satisfies

$$b = e^{\ln b}$$

If b^x is to preserve the logarithm and exponent rules, then one must have

$$b^x = (e^{\ln b})^x = e^{x\cdot\ln b}$$

for each real number x.

This can be used as an alternative definition of the real number power b^x and agrees with the definition given above using rational exponents and continuity. The definition of exponentiation using logarithms is more common in the context of complex numbers, as discussed below.

Real Exponents with Negative Bases

Powers of a positive real number are always positive real numbers. The solution of $x^2 = 4$, however, can be either 2 or –2. The principal value of $4^{1/2}$ is 2, but –2 is also a valid square root. If the definition of exponentiation of real numbers is extended to allow negative results then the result is no longer well behaved.

Neither the logarithm method nor the rational exponent method can be used to define b^r as a real number for a negative real number b and an arbitrary real number r. Indeed, e^r is positive for every real number r, so $\ln(b)$ is not defined as a real number for $b \leq 0$.

The rational exponent method cannot be used for negative values of b because it relies on continuity. The function $f(r) = b^r$ has a unique continuous extension from the rational numbers to the real numbers for each $b > 0$. But when $b < 0$, the function f is not even continuous on the set of rational numbers r for which it is defined.

For example, consider $b = -1$. The nth root of –1 is –1 for every odd natural number n. So if n is an odd positive integer, $(-1)^{(m/n)} = -1$ if m is odd, and $(-1)^{(m/n)} = 1$ if m is even. Thus the set of rational numbers q for which $(-1)^q = 1$ is dense in the rational numbers, as is the set of q for which $(-1)^q = -1$. This means that the function $(-1)^q$ is not continuous at any rational number q where it is defined.

On the other hand, arbitrary complex powers of negative numbers b can be defined by choosing a *complex* logarithm of b.

Chapter

3

Counting Techniques

Some probability problems can be attacked by specifying a sample space $S = \{a_1, a_2, ..., a_n\}$ in which each simple event has probability $\frac{1}{n}$ (i.e. is "equally likely").

Thus, if a compound event A consists of r simple events, then $P(A) = \frac{r}{n}$. To use this approach we need to be able to count the number of events in S and in A, and this can be tricky. We review here some basic ways to count outcomes from "experiments". These approaches should be familiar from high school mathematics.

General Counting Rules

There are two basic rules for counting which can deal with most problems. We phrase the rules in terms of "jobs" which are to be done.

The Addition Rule

Suppose we can do job 1 in ways and job 2 in ways. Then we can do either job 1 or job 2, but not both, in ways. For example, suppose a class has 30 men and 25 women. There are ways the prof. can pick one student to answer a question.

The Multiplication Rule

Suppose we can do job 1 in p ways and an unrelated job 2 in q ways. Then we can do both job 1 and job 2 in $p \times q$ ways.

For example, to ride a bike, you must have the chain on both a front sprocket and a rear sprocket. For a 21 speed bike there are 3 ways to select the front sprocket and 7 ways to select the rear sprocket.

This linkage of OR with addition and AND with multiplication will occur throughout the course, so it is helpful to make this association in your mind. The only problem with applying it is that questions do not always have an AND or an OR in them. You often have to play around with re-wording the question for yourself to discover implied AND's or OR's.

Example: Suppose we pick 2 numbers at random from digits 1, 2, 3, 4, 5 with replacement. (Note: "with replacement" means that after the first number is picked it is "replaced" in the set of numbers, so it could be picked again as the second number.) Let us find the probability that one number is even. This can be reworded as: "The first number is even AND the second is odd, OR, the first is odd AND the second is even." We can then use the addition and multiplication rules to calculate that there are $(2\times3)+(3\times2)=12$ ways for this event to occur. Since the first number can be chosen in 5 ways AND the second in 5 ways, S contains $5\times5=25$ points. The phrase "at random" in the first sentence means the numbers are equally likely to be picked. $Therefore\,P(one\,number\,is\,even)=\frac{12}{25}$

When objects are selected and replaced after each draw, the addition and multiplication rules are generally sufficient to find probabilities. When objects are drawn without being replaced, some special rules may simplify the solution.

Problems:

- A course has 4 sections with no limit on how many can enrol in each section. 3 students each randomly pick a section. Find the probability:
- they all end up in the same section
- they all end up in different sections
- nobody picks section 1.
- Repeat (a) in the case when there are n sections and S students $(n \geq s)$.
- Canadian postal codes consist of 3 letters alternated with 3 digits, starting with a letter (e.g. N2L 3G1). For a randomly constructed postal code, what is the probability:
- all 3 letters are the same?

- the digits are all even or all odd? Treat 0 as being neither even nor odd.
- Suppose a password has to contain between six and eight digits, with each digit either a letter or a number from 1 to 9. There must be at least one number present.
- What is the total number of possible passwords?
- If you started to try passwords in random order, what is the probability you would find the correct password for a given situation within the first 1,000 passwords you tried?

Bayes' Theorem

In probability theory and statistics, Bayes' theorem (alternatively Bayes' law) is a theorem with two distinct interpretations. In the *Bayesian interpretation*, it expresses how a subjective degree of belief should rationally change to account for evidence. In the *frequentist interpretation*, it relates inverse representations of the probabilities concerning two events. In the Bayesian interpretation, Bayes' theorem is fundamental to Bayesian statistics, and has applications in fields including science, engineering, economics (particularly microeconomics), game theory, medicine and law. The application of Bayes' theorem to update beliefs is called Bayesian inference.

Bayes' theorem is named after Thomas Bayes (1701–1761), who first suggested using the theorem to update beliefs. His work was significantly edited and updated by Richard Price before it was posthumously read at the Royal Society. The ideas gained limited exposure until they were independently rediscovered and further developed by Laplace, who first published the modern formulation in his 1812 *Théorieanalytique des probabilités*. Until the second half of the 20th century, the Bayesian interpretation was largely rejected by the mathematics community as unscientific. However, it is now widely accepted. This may have been due to the development of computing, which enabled the successful application of Bayesianism to many complex problems. Sir Harold Jeffreys wrote that Bayes' theorem "is to the theory of probability what Pythagoras's theorem is to geometry".

Introductory Example

Suppose someone told you they had a nice conversation with someone on the train. Not knowing anything else about this conversation, the probability that they were speaking to a woman is 50%. Now suppose they also told you that this person had long hair. It is now more likely they were speaking to a woman, since most long-haired people are women. Bayes' theorem can be used to calculate the probability that the person is a woman.

To see how this is done, let

W represent the event that the conversation was held with a woman, and L denote the event that the conversation was held with a long-haired person.

It can be assumed that women constitute half the population for this example. So, not knowing anything else, the probability that occurs is

$$P(W) = 0.5.$$

Suppose it is also known that 75% of women have long hair, which we denote as

$$P(L \mid W) = 0.75$$

(read: the probability of event L given event W is 0.75).

Likewise, suppose it is known that 30% of men have long hair, or

$$P(L \mid M) = 0.30$$

where M is the complementary event of W, i.e., the event that the conversation was held with a man (assuming that every human is either a man or a woman).

Our goal is to calculate the probability that the conversation was held with a woman, given the fact that the person had long hair, or, in our notation, $P(W \mid L)$. Using the formula for Bayes' theorem, we have:

$$P(W \mid L) = \frac{P(L \mid W)P(W)}{P(L)} = \frac{P(L \mid W)P(W)}{P(L \mid W)P(W) + P(L \mid M)P(M)}$$

where we have used the law of total probability. The numeric answer can be obtained by substituting the above values into this formula. This yields

$$P(W \mid L) = \frac{0.75 \cdot 0.50}{0.75 \cdot 0.50 + 0.30 \cdot 0.50} \approx 0.714,$$

i.e., the probability that the conversation was held with a woman, given that the person had long hair, is about 71%.

Statement and Interpretation

Mathematically, Bayes' theorem gives the relationship between the probabilities of A and B, $P(A)$ and $P(B)$, and the conditional probabilities of given and given , $P(A \mid B)$ and $P(B \mid A)$. In its most common form, it is:

$$P(A \mid B) = \frac{P(B \mid A)P(A)}{P(B)}.$$

The meaning of this statement depends on the interpretation of probability ascribed to the terms:

Bayesian Interpretation

In the Bayesian (or epistemological) interpretation, probability measures a *degree of belief*. Bayes' theorem then links the degree of belief in a proposition before and after accounting for evidence. For example, suppose somebody proposes that a biased coin is twice as likely to land heads than tails. Degree of belief in this might initially be 50%. The coin is then flipped a number of times to collect evidence. Belief may rise to 70% if the evidence supports the proposition.

For proposition A and evidence B,

- $P(A)$, the *prior*, is the initial degree of belief in A.
- $P(A \mid B)$, the *posterior*, is the degree of belief having accounted for B.
- represents the support B provides for A.

For more on the application of Bayes' theorem under the Bayesian interpretation of probability.

Frequentist Interpretation

In the frequentist interpretation, probability measures a *proportion of outcomes*. For example, suppose an experiment is performed many times. $P(A)$ is the proportion of outcomes with property A, and $P(B)$ that with property B. $P(B \mid A)$ is the proportion of outcomes with property *out of* outcomes with property A, and the $P(A \mid B)$ proportion of those with out of those with B.

The role of Bayes' theorem is best visualized with tree diagrams. The two diagrams partition the same outcomes by A and B in opposite orders, to obtain the inverse probabilities. Bayes' theorem serves as the link between these different partitionings.

Forms

Events:

Simple form

For events A and B, provided that $P(B) \neq 0$,

$$P(A \mid B) = \frac{P(B \mid A)P(A)}{P(B)}.$$

In a Bayesian inference step, the probability of evidence B is constant for all models A_n. The posterior may then be expressed as proportional to the numerator:

$$P(A_n \mid B) \propto P(B \mid A_n)P(A_n).$$

Extended form

Often, for some partition $\{A_i\}$ of the event space, the event space is given or conceptualized in terms of $P(A_i)$ and $P(B \mid A_i)$. It is then useful to eliminate $P(B)$ using the law of total probability:

$$P(B) = \sum_j P(B \mid A_j)P(A_j),$$

$$\Rightarrow P(A_i \mid B) = \frac{P(B \mid A_i)P(A_i)}{\sum_j P(B \mid A_j)P(A_j)}.$$

In the special case of a binary partition,

$$P(A \mid B) = \frac{P(B \mid A)P(A)}{P(B \mid A)P(A) + P(B \mid \neg A)P(\neg A)}.$$

Random Variables

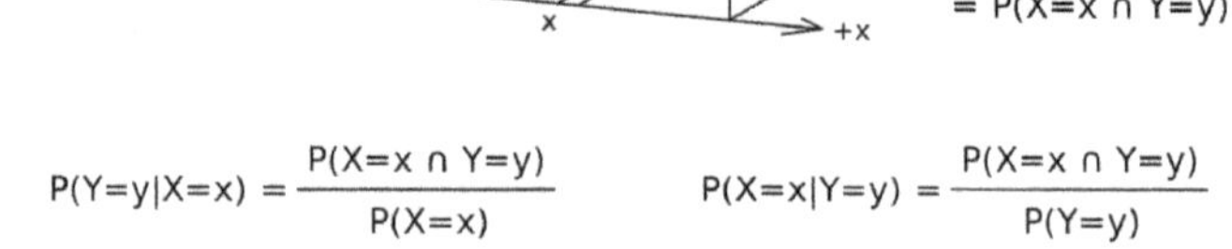

Figure: *Diagram illustrating the meaning of Bayes' theorem as applied to an event space generated by continuous random variables X and Y. Note that there exists an instance of Bayes' theorem for each point in the domain. In practise, these instances might be parametrised by writing the specified probability densities as a function of x and y.*

Consider a sample space Ω generated by two random variables X and Y. In principle, Bayes' theorem applies to the events $A = \{X = x\}$ and $B = \{Y = y\}$. However, terms become 0 at points where either variable has finite probability density. To remain useful, Bayes' theorem may be formulated in terms of the relevant densities.

Simple form

If X is continuous and Y is discrete,

$$f_X(x \mid Y=y) = \frac{P(Y=y \mid X=x) f_X(x)}{P(Y=y)}.$$

If X is discrete and Y is continuous,

$$P(X=x \mid Y=y) = \frac{f_Y(y \mid X=x) P(X=x)}{f_Y(y)}.$$

If both X and Y are continuous,

$$f_X(x \mid Y=y) = \frac{f_Y(y \mid X=x) f_X(x)}{f_Y(y)}.$$

Extended form

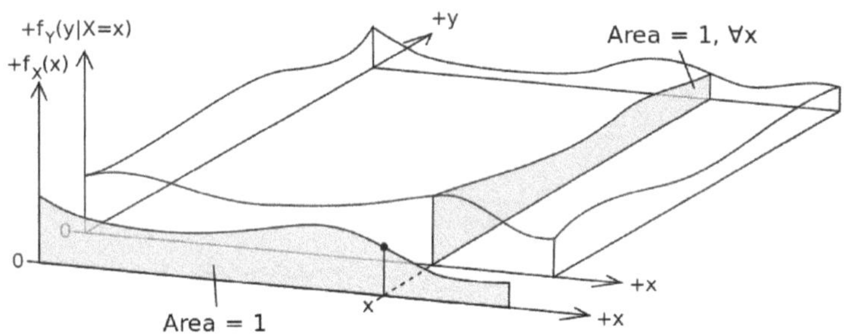

Figure: *Diagram illustrating how an event space generated by continuous random variables X and Y is often conceptualized.*

A continuous event space is often conceptualized in terms of the numerator terms. It is then useful to eliminate the denominator using the law of total probability. For $f_Y(y)$, this becomes an integral:

$$f_Y(y) = \int_{-\infty}^{\infty} f_Y(y \mid X=\xi) f_X(\xi) d\xi.$$

Bayes' rule

Bayes' rule is Bayes' theorem in odds form.

$$O(A_1 : A_2 \mid B) = O(A_1 : A_2) \cdot \Lambda(A_1 : A_2 \mid B)$$

where

$$\Lambda(A_1 : A_2 \mid B) = \frac{P(B \mid A_1)}{P(B \mid A_2)}$$

So the rule says that the posterior odds are the prior odds times the Bayes factor.

Derivation

For Events: Bayes' theorem may be derived from the definition of conditional probability:

$$P(A \mid B) = \frac{P(A \cap B)}{P(B)}, \text{ if } P(B) \neq 0,$$

$$P(B \mid A) = \frac{P(A \cap B)}{P(A)}, \text{ if } P(A) \neq 0,$$

$$\Rightarrow P(A \cap B) = P(A \mid B)P(B) = P(B \mid A)P(A),$$

$$\Rightarrow P(A \mid B) = \frac{P(B \mid A)P(A)}{P(B)}, \text{ if } P(B) \neq 0.$$

For Random Variables

For two continuous random variables X and Y, Bayes' theorem may be analogously derived from the definition of conditional density:

$$f_X(x \mid Y = y) = \frac{f_{X,Y}(x,y)}{f_Y(y)}$$

$$f_Y(y \mid X = x) = \frac{f_{X,Y}(x,y)}{f_X(x)}$$

$$\Rightarrow f_X(x \mid Y = y) = \frac{f_Y(y \mid X = x) f_X(x)}{f_Y(y)}.$$

Examples: Frequentist example

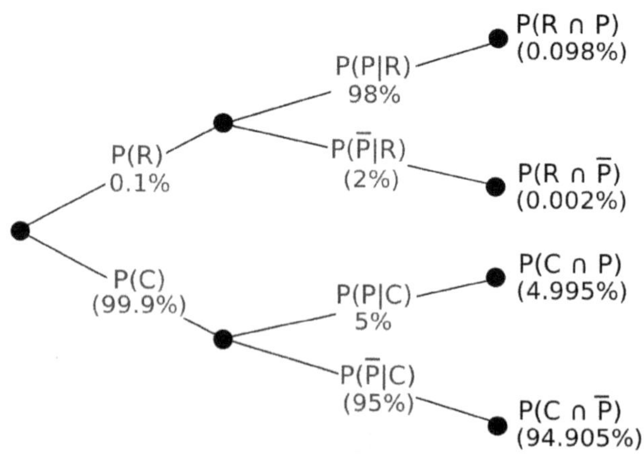

Figure: *Tree diagram illustrating frequentist example. R, C, P and P bar are the events representing rare, common, pattern and no pattern. Percentages in parentheses are calculated. Note that three independent values are given, so it is possible to calculate the inverse tree.*

An entomologist spots what might be a rare subspecies of beetle, due to the pattern on its back. In the rare subspecies, 98% have the pattern. In the common subspecies, 5% have the pattern. The rare subspecies accounts for only 0.1% of the population. How likely is the beetle to be rare?

From the extended form of Bayes' theorem,

$$\begin{aligned} P(\text{Rare} \mid \text{Pattern}) &= \frac{P(\text{Pattern} \mid \text{Rare})P(\text{Rare})}{P(\text{Pattern} \mid \text{Rare})P(\text{Rare}) + P(\text{Pattern} \mid \text{Common})P(\text{Common})} \\ &= \frac{0.98 \times 0.001}{0.98 \times 0.001 + 0.05 \times 0.999} \\ &\approx 1.9\%. \end{aligned}$$

Drug Testing

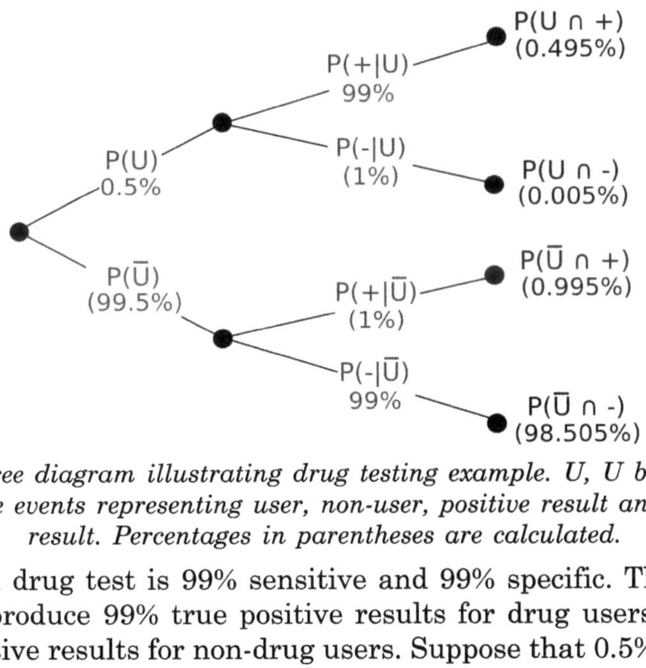

Figure: *Tree diagram illustrating drug testing example. U, U bar, "+" and "–" are the events representing user, non-user, positive result and negative result. Percentages in parentheses are calculated.*

Suppose a drug test is 99% sensitive and 99% specific. That is, the test will produce 99% true positive results for drug users and 99% true negative results for non-drug users. Suppose that 0.5% of people are users of the drug. If a randomly selected individual tests positive, what is the probability he or she is a user?

$$\begin{aligned} P(\text{User} \mid +) &= \frac{P(+ \mid \text{User})P(\text{User})}{P(+ \mid \text{User})P(\text{User}) + P(+ \mid \text{Non-user})P(\text{Non-user})} \\ &= \frac{0.99 \times 0.005}{0.99 \times 0.005 + 0.01 \times 0.995} \\ &\approx 33.2\% \end{aligned}$$

Despite the apparent accuracy of the test, if an individual tests positive, it is more likely that they do *not* use the drug than that they do.

This surprising result arises because the number of non-users is very large compared to the number of users, such that the number of false positives (0.995%) outweighs the number of true positives (0.495%). To use concrete numbers, if 1000 individuals are tested, there are expected to be 995 non-users and 5 users. From the 995 non-users, $0.01 \times 995 \approx 10$ false positives are expected. From the 5 users, $0.99 \times 5 \approx 5$ true positives are expected. Out of 15 positive results, only 5, about 33%, are genuine.

History

Bayes' theorem was named after the Reverend Thomas Bayes (1701–61), who studied how to compute a distribution for the probability parameter of a binomial distribution (in modern terminology). His friend Richard Price edited and presented this work in 1763, after Bayes' death, as *An Essay towards solving a Problem in the Doctrine of Chances*. The French mathematician Pierre-Simon Laplace reproduced and extended Bayes' results in 1774, apparently quite unaware of Bayes' work. Stephen Stigler suggested in 1983 that Bayes' theorem was discovered by Nicholas Saunderson some time before Bayes. However, this interpretation has been disputed.

Stephen Fienberg describes the evolution from "inverse probability" at the time of Bayes and Laplace (a term still used by Harold Jeffreys in 1939), to "Bayesian" in the 1950s and notes that Ronald A. Fisher was the first to use the term "Bayesian" (albeit in a derogatory sense).

Random Variable

In probability and statistics, a random variable or stochastic variable is a variable whose value is subject to variations due to chance (i.e. randomness, in a mathematical sense). As opposed to other mathematical variables, a random variable conceptually does not have a single, fixed value (even if unknown); rather, it can take on a set of possible different values, each with an associated probability.

A random variable's possible values might represent the possible outcomes of a yet-to-be-performed experiment or an event that has not happened yet, or the potential values of a past experiment or event whose already-existing value is uncertain (e.g. as a result of incomplete information or imprecise measurements). They may also conceptually represent either the results of an "objectively" random process (e.g. rolling a die), or the "subjective" randomness that results from incomplete knowledge of a quantity. The meaning of the probabilities assigned to the potential values of a random variable is not part of probability theory itself, but instead related to philosophical arguments over the interpretation of probability. The mathematics works the same regardless of the particular interpretation in use.

Random variables can be classified as either *discrete* (i.e. it may assume any of a specified list of exact values) or as *continuous* (i.e. it may assume any numerical value in an interval or collection of intervals). The mathematical function describing the possible values of a random variable and their associated probabilities is known as a probability distribution. The realizations of a random variable, i.e. the results of randomly choosing values according to the variable's probability distribution are called random variates.

The basic concept of "random variable" in statistics is real-valued. However, one can consider arbitrary types such as boolean values, categorical variables, complex numbers, vectors, matrices, sequences, trees, sets, shapes, manifolds, functions, and processes. The term *random element* is used to encompass all such related concepts. An example is the stochastic process, a set of indexed random variables (typically indexed by time or space). These more general concepts are particularly useful in fields such as computer science and natural language processing where many of the basic elements of analysis are non-numerical. Such general random elements can sometimes be treated as sets of real-valued random variables — often more specifically as random vectors. For example:

- A "random word" may be parameterized by an integer-valued index into the vocabulary of possible words; or alternatively as an indicator vector, in which exactly one element is a 1 and the others are 0, with the 1 indexing a particular word into a vocabulary.
- A "random sentence" may be parameterized as a vector of random words.
- A random graph, for a graph with V edges, may be parameterized as an NxN matrix, indicating the weight for each edge, or 0 for no edge. (If the graph has no weights, 1 indicates an edge, 0 indicates no edge.)

However, reduction to numerical values is not essential for dealing with random elements: a randomly selected individual remains an individual, not a number.

The formal mathematical treatment of random variables is dealt with in the subject of probability theory. In that context, random variables are defined in terms of functions defined on a probability space.

Definition

A random variable is defined on a set of possible outcomes (the sample space Ω) and a probability distribution that associates each outcome with a probability. A random variable represents a measurable aspect or property of the outcomes, and hence associates each outcome with a number.

That is, a random variable is formally defined as a mapping (a function) from the sample space Ω to the real numbers. In an experiment a person may be chosen at random, and one random variable may be its age, and another its number of children. Random variables are typically classified as either discrete or continuous. Discrete variables can take on either a finite or at most a countably infinite set of discrete values. Their probability distribution is given by a probability mass function which directly maps a value of the random variable to a probability. Continuous variables, however, take on values that vary continuously within one or more (possibly infinite) intervals.

As a result there are an uncountably infinite number of individual outcomes, and each has a probability 0. As a result, the probability distribution for many continuous random variables is defined using a probability density function, which indicates the "density" of probability in a small neighbourhood around a given value. More technically, the probability that an outcome is in a particular range is derived from the integration of the probability density function in that range.

In the case where the sample space is some subset of the real numbers, both concepts can be united using a cumulative distribution function (CDF), which describes the probability that an outcome will be less than or equal to a specified value. This function is necessarily monotonically non-decreasing, with a minimum value of 0 at negative infinity and a maximum value of 1 at positive infinity. The CDF of a discrete distribution will consist mostly of flat areas along with sudden jumps at each outcome defined in the sample space, while the CDF of a continuous distribution will typically rise gradually and continuously. Distributions that are partly discrete and partly continuous can also be described this way.

Examples

The possible outcomes for one coin toss can be described by the sample space Ω= {heads, tails}. We can introduce a real-valued random variable Y that models a \$1 payoff for a successful bet on heads as follows:

$$Y(\omega)=\begin{cases}1, & \text{if } \omega=\text{heads},\\ 0, & \text{if } \omega=\text{tails}.\end{cases}$$

If the coin is equally likely to land on either side then Y has a probability mass function given by:

$$\rho_Y(y)=\begin{cases}\frac{1}{2}, & \text{if } y=1,\\ \frac{1}{2}, & \text{if } y=0,\\ 0, & \text{otherwise.}\end{cases}$$

A random variable can also be used to describe the process of rolling a die and the possible outcomes. The most obvious representation is to take the set $\Omega=\{1, 2, 3, 4, 5, 6\}$ as the sample space, defining the random variable X to be equal to the number rolled. In this case,

$$X(\omega)=\omega$$

and

$$\rho_X(x)=\begin{cases}\frac{1}{6}, & \text{if } x=1,2,3,4,5,6,\\ 0, & \text{otherwise.}\end{cases}$$

An example of a continuous random variable would be one based on a spinner that can choose a horizontal direction. Then the values taken by the random variable are directions. We could represent these directions by North, West, East, South, Southeast, etc. However, it is commonly more convenient to map the sample space to a random variable which takes values which are real numbers. This can be done, for example, by mapping a direction to a bearing in degrees clockwise from North.

The random variable then takes values which are real numbers from the interval [0, 360), with all parts of the range being "equally likely". In this case, $X=$ the angle spun. Any real number has probability zero of being selected, but a positive probability can be assigned to any *range* of values. For example, the probability of choosing a number in [0, 180] is ½. Instead of speaking of a probability mass function, we say that the probability *density* of X is 1/360. The probability of a subset of [0, 360) can be calculated by multiplying the measure of the set by 1/360. In general, the probability of a set for a given continuous random variable can be calculated by integrating the density over the given set.

An example of a random variable of mixed type would be based on an experiment where a coin is flipped and the spinner is spun only if the result of the coin toss is heads. If the result is tails, $X = -1$; otherwise X = the value of the spinner as in the preceding example. There is a probability of ½ that this random variable will have the value −1. Other ranges of values would have half the probability of the last example.

Distribution Functions of Random Variables

If a random variable $X : \Omega \to \mathbb{R}$ defined on the probability space $(\Omega, \mathcal{F}, P)$ is given, we can ask questions like "How likely is it that the value of X is bigger than 2?". This is the same as the probability of the event $\{\omega : X(\omega) > 2\}$ which is often written as $P(X > 2)$ for short.

Recording all these probabilities of output ranges of a real-valued random variable X yields the probability distribution of X. The probability distribution "forgets" about the particular probability space used to define X and only records the probabilities of various values of X. Such a probability distribution can always be captured by its cumulative distribution function

$$F_X(x) = \mathrm{P}(X \le x)$$

and sometimes also using a probability density function. In measure-theoretic terms, we use the random variable X to "push-forward" the measure P on Ω to a measure dF on R. The underlying probability space Ω is a technical device used to guarantee the existence of random variables, sometimes to construct them, and to define notions such as correlation and dependence or independence based on a joint distribution of two or more random variables on the same probability space. In practice, one often disposes of the space Ω altogether and just

puts a measure on R that assigns measure 1 to the whole real line, i.e., one works with probability distributions instead of random variables.

Moments

The probability distribution of a random variable is often characterised by a small number of parameters, which also have a practical interpretation. For example, it is often enough to know what its "average value" is. This is captured by the mathematical concept of expected value of a random variable, denoted E[X], and also called the first moment. In general, E[$f(X)$] is not equal to f(E[X]). Once the "average value" is known, one could then ask how far from this average value the values of X typically are, a question that is answered by the variance and standard deviation of a random variable. E[X] can be viewed intuitively as an average obtained from an infinite population, the members of which are particular evaluations of X.

Mathematically, this is known as the (generalised) problem of moments: for a given class of random variables X, find a collection $\{f_i\}$ of functions such that the expectation values E[$f_i(X)$] fully characterise the distribution of the random variable X.

Moments can only be defined for real-valued functions of random variables. If the random variable is itself real-valued, then moments of the variable itself can be taken, which are equivalent to moments of the identity function $f(X) = X$ of the random variable. However, even for non-real-valued random variables, moments can be taken of real-valued functions of those variables. For example, for a categorical random variable X that can take on the nominal values "red", "blue" or "green", the real-valued function $[X = \text{green}]$ can be constructed; this uses the I verson bracket, and has the value 1 if X has the value "green", 0 otherwise. Then, the expected value and other moments of this function can be determined.

Functions of Random Variables

A new random variable Y can be defined by applying a real Borel measurable function $g : \mathbb{R} \to \mathbb{R}$ to the outcomes of a real-valued random variable X. The cumulative distribution function of Y is

$$F_Y(y) = P(g(X) \le y).$$

If function g is invertible, i.e. g^{-1} exists, and increasing, then the previous relation can be extended to obtain

$$F_Y(y) = P(g(X) \le y) = P(X \le g^{-1}(y)) = F_X(g^{-1}(y))$$

and, again with the same hypotheses of invertibility of g, assuming also differentiability, we can find the relation between the probability density functions by differentiating both sides with respect to y, in order to obtain

$$f_Y(y) = f_X(g^{-1}(y))\left|\frac{dg^{-1}(y)}{dy}\right|.$$

If there is no invertibility of g but each y admits at most a countable number of roots (i.e. a finite, or countably infinite, number of x_i such that $y = g(x_i)$) then the previous relation between the probability density functions can be generalized with

$$f_Y(y) = \sum_i f_X(g_i^{-1}(y)) \left| \frac{dg_i^{-1}(y)}{dy} \right|$$

where $x_i = g_i^{-1}(y)$. The formulas for densities do not demand g to be increasing.

In the measure-theoretic, axiomatic approach to probability, if we have a random variable X on $\Omega$and a Borel measurable function $g : \mathbb{R} \to \mathbb{R}$, then $Y = g(X)$ will also be a random variable on Ω, since the composition of measurable functions is also measurable. (However, this is not true if is g Lebesgue measurable.) The same procedure that allowed one to go from a probability space to can be used (Ω,P) to $(\mathbb{R},dF_X)$ obtain the distribution of Y.

Equivalence of Random Variables

There are several different senses in which random variables can be considered to be equivalent. Two random variables can be equal, equal almost surely, or equal in distribution.

In increasing order of strength, the precise definition of these notions of equivalence is given below.

Equality in Distribution

If the sample space is a subset of the real line a possible definition is that random variables X and Y are *equal in distribution* if they have the same distribution functions:

$$P(X \le x) = P(Y \le x) \quad \textit{for all} \quad x.$$

Two random variables having equal moment generating functions have the same distribution. This provides, for example, a useful method of checking equality of certain functions of i.i.d. random variables. However, the moment generating function exists only for distributions that have a defined Laplace transform.

Almost Sure Equality

Two random variables X and Y are *equal almost surely* if, and only if, the probability that they are different is zero:

$$P(X \ne Y) = 0.$$

For all practical purposes in probability theory, this notion of equivalence is as strong as actual equality. It is associated to the following distance:

$$d_\infty(X,Y) = \operatorname*{ess\,sup}_{\omega} |X(\omega) - Y(\omega)|,$$

where "ess sup" represents the essential supremum in the sense of measure theory.

Equality

Finally, the two random variables X and Y are *equal* if they are equal as functions on their measurable space:

$$X(\omega) = Y(\omega) \qquad \text{for all } \omega.$$

Convergence

A significant theme in mathematical statistics consists of obtaining convergence results for certain sequences of random variables; for instance the law of large numbers and the central limit theorem. There are various senses in which a sequence (X_n) of random variables can converge to a random variable X. These are explained in the article on convergence of random variables.

Mean

In statistics, mean has three related meanings:

- the arithmetic mean of a sample (distinguished from the geometric mean or harmonic mean).
- the expected value of a random variable.
- the mean of a probability distribution.

There are other statistical measures of central tendency that should not be confused with means - including the 'median' and 'mode'. Statistical analyses also commonly use measures of dispersion, such as the range, interquartile range, or standard deviation. Note that not every probability distribution has a defined mean.

For a data set, the arithmetic mean is equal to the sum of the values divided by the number of values. The arithmetic mean of a set of numbers x_1, x_2, ..., x_n is typically denoted by $\bar{x}$, pronounced "x bar". If the data set were based on a series of observations obtained by sampling from a statistical population, the arithmetic mean is termed the "sample mean" ($\bar{x}$) to distinguish it from the "population mean" (μ or μ_x). For a finite population, the population mean of a property is equal to the arithmetic mean of the given property while considering every member of the population. For example, the population mean height is equal to the sum of the heights of every individual divided by the total number of individuals.

The sample mean may differ from the population mean, especially for small samples. The law of large numbers dictates that the larger the size of the sample, the more likely it is that the sample mean will be close to the population mean.

For a probability distribution, the mean is equal to the sum or integral over every possible value weighted by the probability of that value. In the case of a discrete probability distribution, the mean of a discrete random variable x is computed by taking the product of each possible value of x and its probability $P(x)$, and then adding all these products together, giving $\mu = \sum xP(x)$.

As well as statistics, means are often used in geometry and analysis; a wide range of means have been developed for these purposes, which are not much used in statistics. Examples of means are listed below.

Examples of Means

Arithmetic Mean (AM)

The *arithmetic mean* is the "standard" average, often simply called the "mean".

$$\bar{x} = \frac{1}{n} \cdot \sum_{i=1}^{n} x_i$$

For example, the arithmetic mean of five values: 4, 36, 45, 50, 75 is

$$\frac{4+36+45+50+75}{5} = \frac{210}{5} = 42.$$

The mean may often be confused with the median, mode or range. The mean is the arithmetic average of a set of values, or distribution; however, for skewed distributions, the mean is not necessarily the same as the middle value (median), or the most likely (mode). For example, mean income is skewed upwards by a small number of people with very large incomes, so that the majority have an income lower than the mean. By contrast, the median income is the level at which half the population is below and half is above. The mode income is the most likely income, and favours the larger number of people with lower incomes. The median or mode are often more intuitive measures of such data. Nevertheless, many skewed distributions are best described by their mean – such as the exponential and Poisson distributions.

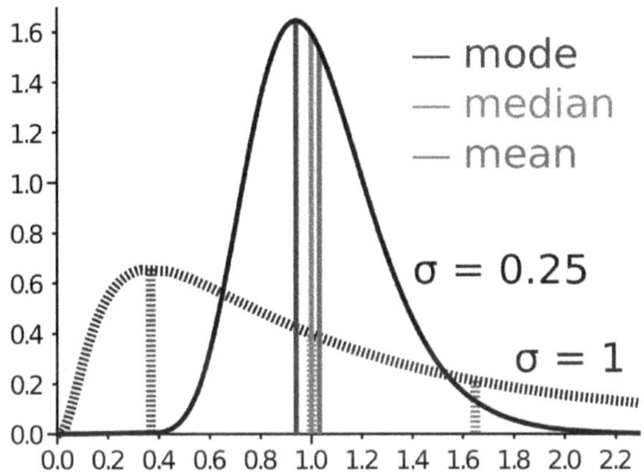

***Figure:** Comparison of the arithmetic mean, median and mode of two skewed (log-normal) distributions.*

Geometric Mean (GM)

The geometric mean is an average that is useful for sets of positive numbers that are interpreted according to their product and not their sum (as is the case with the arithmetic mean) e.g. rates of growth.

$$\bar{x} = \left(\prod_{i=1}^{n} x_i \right)^{\frac{1}{n}}$$

For example, the geometric mean of five values: 4, 36, 45, 50, 75 is:

$$(4\times 36\times 45\times 50\times 75)^{1/_5} = \sqrt[5]{24\,300\,000} = 30.$$

Harmonic Mean (HM)

The harmonic mean is an average which is useful for sets of numbers which are defined in relation to some unit, for example speed (distance per unit of time).

$$\bar{x} = n\cdot\left(\sum_{i=1}^{n} \frac{1}{x_i} \right)^{-1}$$

For example, the harmonic mean of the five values: 4, 36, 45, 50, 75 is

$$\frac{5}{\frac{1}{4}+\frac{1}{36}+\frac{1}{45}+\frac{1}{50}+\frac{1}{75}} = \frac{5}{\frac{1}{3}} = 15.$$

Relationship between AM, GM, and HM

AM, GM, and HM satisfy these inequalities:

$$AM \geq GM \geq HM$$

Equality holds only when all the elements of the given sample are equal.

Generalized Means

Power Mean: The generalized mean, also known as the power mean or Hölder mean, is an abstraction of the quadratic, arithmetic, geometric and harmonic means. It is defined for a set of n positive numbers x_i by

$$\bar{x}(m) = \left(\frac{1}{n} \cdot \sum_{i=1}^{n} x_i^m\right)^{\frac{1}{m}}$$

By choosing different values for the parameter m, the following types of means are obtained:

$m \to \infty$ maximum

$m = 2$ quadratic mean

$m = 1$ arithmetic mean

$m \to 0$ geometric mean

$m = -1$ harmonic mean

$m \to -\infty$ minimum

f-mean

This can be generalized further as the generalized f-mean

$$\bar{x} = f^{-1}\left(\frac{1}{n} \cdot \sum_{i=1}^{n} f(x_i)\right)$$

and again a suitable choice of an invertible f will give

$f(x) = x$ arithmetic mean,

$f(x) = \frac{1}{x}$ harmonic mean,

$f(x) = x^m$ power mean,

$f(x) = \ln x$ geometric mean.

Weighted Arithmetic Mean

The weighted arithmetic mean (or weighted average) is used, if one wants to combine average values from samples of the same population with different sample sizes:

$$\overline{x} = \frac{\sum_{i=1}^{n} w_i \cdot x_i}{\sum_{i=1}^{n} w_i}.$$

The weights w_i represent the bounds of the partial sample. In other applications they represent a measure for the reliability of the influence upon the mean by respective values.

Truncated mean

Sometimes a set of numbers might contain outliers, i.e., a datum which is much lower or much higher than the others. Often, outliers are erroneous data caused by artifacts. In this case, one can use a truncated mean.

It involves discarding given parts of the data at the top or the bottom end, typically an equal amount at each end, and then taking the arithmetic mean of the remaining data. The number of values removed is indicated as a percentage of total number of values.

Interquartile Mean

The interquartile mean is a specific example of a truncated mean. It is simply the arithmetic mean after removing the lowest and the highest quarter of values.

$$\overline{x} = \frac{2}{n} \sum_{i=(n/4)+1}^{3n/4} x_i$$

assuming the values have been ordered, so is simply a specific example of a weighted mean for a specific set of weights.

Mean of a Function

In calculus, and especially multivariable calculus, the mean of a function is loosely defined as the average value of the function over its domain. In one variable, the mean of a function *f*(*x*) over the interval (*a*,*b*) is defined by

$$\overline{f} = \frac{1}{b-a} \int_a^b f(x)dx.$$

Recall that a defining property of the average value $\overline{y}$ of finitely many numbers $y_1, y_2, \ldots, y_n$ is that $n\overline{y} = y_1 + y_2 + \cdots + y_n$. In other words, $\overline{y}$ is the *constant* value which when *added* to itself *n* times equals the result of adding the n terms of y_i. By analogy, a defining property $\bar{f}$ of the average value of a function over the interval $[a,b]$ is that

$$\int_a^b \bar{f}\,dx = \int_a^b f(x)dx$$

In other words, $\bar{f}$ is the *constant* value which when *integrated* over $[a,b]$ equals the result of integrating $f(x)$ over $[a,b]$. But by the second fundamental theorem of calculus, the integral of a constant is just

$$\int_a^b \bar{f}\,dx = \bar{f}x\,\Big|_a^b = \bar{f}b - \bar{f}a = (b-a)\bar{f}$$

See also the first mean value theorem for integration, which guarantees that if f is continuous then there exists a point $c \in (a,b)$ such that

$$\int_a^b f(x)dx = f(c)(b-a)$$

The point $f(c)$ is called the mean value of $f(x)$ on $[a,b]$. So we write $\bar{f} = f(c)$ and rearrange the preceding equation to get the above definition.

In several variables, the mean over a relatively compact domain U in a Euclidean space is defined by

$$\bar{f} = \frac{1}{\mathrm{Vol}(U)}\int_U f.$$

This generalizes the arithmetic mean. On the other hand, it is also possible to generalize the geometric mean to functions by defining the geometric mean of f to be

$$\exp\left(\frac{1}{\mathrm{Vol}(U)}\int_U \log f\right).$$

More generally, in measure theory and probability theory, either sort of mean plays an important role. In this context, Jensen's inequality places sharp estimates on the relationship between these two different notions of the mean of a function.

There is also a *harmonic average* of functions and a *quadratic average* (or *root mean square*) of functions.

Mean of a Probability Distribution

Mean of Angles: Most of the usual means fail on circular quantities, like angles, daytimes, fractional parts of real numbers. For those quantities you need a mean of circular quantities.

Fréchet Mean

The Fréchet mean gives a manner for determining the "centre" of a mass distribution on a surface or, more generally, Riemannian manifold.

Unlike many other means, the Fréchet mean is defined on a space whose elements cannot necessarily be added together or multiplied by scalars. It is sometimes also known as the Karcher mean (named after Hermann Karcher).

Properties

All means share some properties and additional properties are shared by the most common means. Some of these properties are collected here.

Weighted Mean

A weighted mean M is a function which maps tuples of positive numbers to a positive number

$$M : (0,\infty)^n \to (0,\infty)$$

such that the following properties hold:

- *"Fixed point":* $M(1,1,...,1) = 1$
- *Homogeneity:* $M(\lambda\, x_1, ..., \lambda\, x_n) = \lambda\, M(x_1, ..., x_n)$ for all λ *and* x_i. *In vector notation:* $M(\lambda\, x) = \lambda\, Mx$ for all n-vectors x.
- *Monotonicity:* If $x_i \le y_i$ for each i, then $Mx \le My$

It follows

- *Boundedness:* $\min x \le Mx \le \max x$
- *Continuity:* $\lim_{x\to y} Mx = My$
- There are means which are not differentiable. For instance, the maximum number of a tuple is considered a mean (as an extreme case of the power mean, or as a special case of a median), but is not differentiable.
- All means listed above, with the exception of most of the Generalized f-means, satisfy the presented properties.
  * If f is bijective, then the generalized f-mean satisfies the fixed point property.
  * If f is strictly monotonic, then the generalized f-mean satisfy also the monotony property.
  * In general a generalized f-mean will miss homogeneity.

The above properties imply techniques to construct more complex means:

If C, M_1, ..., M_m are weighted means and p is a positive real number, then A and B defined by

$$Ax = C(M_1 x, \ldots, M_m x),$$

$$Bx = \sqrt[p]{C(x_1^p, \ldots, x_n^p)},$$

are also weighted means.

Unweighted Mean

Intuitively spoken, an unweighted mean is a weighted mean with equal weights. Since our definition of *weighted mean* above does not expose particular weights, equal weights must be asserted by a different way. A different view on homogeneous weighting is, that the inputs can be swapped without altering the result.

Thus we define M to be an unweighted mean if it is a weighted mean and for each permutation π of inputs, the result is the same.

Symmetry: $Mx = M(\pi x)$ for all n-tuples x and permutations π on n-tuples.

Analogously to the weighted means, if C is a weighted mean and M_1, ..., M_m are unweighted means and p is a positive real number, then A and B defined by

$$Ax = C(M_1 x, \ldots, M_m x),$$

$$Bx = \sqrt[p]{M_1(x_1^p, \ldots, x_n^p)},$$

are also unweighted means.

Convert Unweighted Mean to Weighted Mean

An unweighted mean can be turned into a weighted mean by repeating elements. This connection can also be used to state that a mean is the weighted version of an unweighted mean. Say you have the unweighted mean M and weight the numbers by natural numbers $a_1, \ldots, a_n$. (If the numbers are rational, then multiply them with the least common denominator.)

Then the corresponding weighted mean A is obtained by

$$A(x_1, \ldots, x_n) = M(\underbrace{x_1, \ldots, x_1}_{a_1}, x_2, \ldots, x_{n-1}, \underbrace{x_n, \ldots, x_n}_{a_n}).$$

Means of Tuples of Different Sizes

If a mean M is defined for tuples of several sizes, then one also expects that the mean of a tuple is bounded by the means of partitions. More precisely

- Given an arbitrary tuple x, which is partitioned into $y_1, ..., y_k$, then

$$Mx \in \text{convexhull}(My_1, \ldots, My_k).$$

Population and Sample Means

The mean of a population is denoted μ, known as the population mean. The sample mean makes a good estimator of the population mean, as its expected value is equal to the population mean. The sample mean of a population is a random variable, not a constant, and consequently it will have its own distribution. For a random sample of n observations from a normally distributed population, the sample mean distribution is

$$\bar{x} \sim N\left\{\mu, \frac{\sigma^2}{n}\right\}.$$

Often, since the population variance is an unknown parameter, it is estimated by the mean sum of squares, which changes the distribution of the sample mean from a normal distribution to a Student's t distribution with $n - 1$ degrees of freedom.

Small Sample Sizes

Small sample sizes occur in practice and present unusually difficult problems for parameter estimation.

It is a common but false belief that from a single ($n = 1$) observation x that information about the variability in the population cannot be gained and that consequently that finite-length confidence intervals for mean and/or variance are impossible even in principle. Where the shape of the population distribution is known, some estimates are possible:

For a normally distributed variate, the confidence intervals for the (arithmetic) population mean at the 90% level have been shown to be $x \pm 5.84 \mid x \mid$ where $\mid \mid$ is the absolute value. The 95% bound for a normally distributed variate is $x \pm 9.68 \mid x \mid$ and that for a 99% confidence interval is $x \pm 48.39 \mid x \mid$.

The estimate derived from this method shows behaviour that is atypical of more conventional methods. A value of 0 for the sample mean cannot be rejected with any level of confidence. If $x = 0$, the confidence interval collapses to a length of 0. Finally the confidence interval is not stable under a linear transform $x \rightarrow ax + b$ where a and b are constants.

Machol has shown that given a known density symmetrical about 0 and a single sample value (x) that the 90% confidence intervals of population mean are

$$x \pm 5 | x - \nu |$$

where ν is the population median.

For a sample size of two ($n = 2$), the population mean is bounded by

$$\frac{x_1 + x_2}{2} - k\frac{| x_1 + x_2 |}{2} \leq \mu \frac{x_1 + x_2}{2} + k\frac{| x_1 + x_2 |}{2}$$

where x_1, x_2 are the variate values, μ is the population mean and k is a constant that depends on the underlying distribution. For the normal distribution, k = cotangent($\pi\alpha/2$). For $\alpha = 0.05$, $k = 12.71$. For the rectangular distribution, $k = (1/\alpha) - 1$. For $\alpha = 0.05$, $k = 19$.

For a sample size of three ($n = 3$), the confidence intervals for the population mean are

$$m - k\frac{s}{\sqrt{3}} \leq \mu \leq m + k\frac{s}{\sqrt{3}}$$

where m is the sample mean, s is the sample standard deviation and k is a constant that depends on the distribution. For the normal distribution, k is approximately $1/\sqrt{\alpha} - 3\sqrt{\alpha}/4 + \ldots$ When $\alpha = 0.05$, $k = 4.30$. For the rectangular distribution with $\alpha = 0.05$, $k = 5.74$.

For medium-size samples ($4 \leq n \leq 20$), the pivot statistics give better estimates of the population mean than the t test does.

The pivot depth (j) is int $((n+1)/2)/2$ or int $((n+1)/2+1)/2$ depending on which value is an integer. The lower pivot is $x_L = x_j$ and the upper pivot x_U is x_{n+1-j}. The pivot half sum (P) is

$$P = \frac{x_U + x_L}{2}$$

and the pivot range (R) is

$$R = x_U - x_L$$

The confidence intervals for the population mean are then

$$P - Rt \leq \mu \leq P - Rt$$

where t is the value of the t test at 100($1 - \alpha/2$)%.

The pivot statistic $T = P / R$ has an approximately symmetrical distribution and its values for $4 \leq n \leq 20$ for a number of values of $1-\alpha$ are given in Table of Meloun *et al.*

Median

In statistics and probability theory, the median is the numerical value separating the higher half of a data sample, a population, or a probability distribution, from the lower half. The *median* of a finite list of numbers can be found by arranging all the observations from lowest value to highest value and picking the middle one (eg, the median of {3, 5, 9} is 5). If there is an even number of observations, then there is no single middle value; the median is then usually defined to be the mean of the two middle values.

A median is only defined on ordered one-dimensional data, and is independent of any distance metric. A geometric median, on the other hand, is defined in any number of dimensions.

In a sample of data, or a finite population, there may be no member of the sample whose value is identical to the median (in the case of an even sample size); if there is such a member, there may be more than one so that the median may not uniquely identify a sample member. Nonetheless, the value of the median is uniquely determined with the usual definition. A related concept, in which the outcome is forced to correspond to a member of the sample, is the medoid. At most, half the population have values strictly less than the *median*, and, at most, half have values strictly greater than the median. If each group contains less than half the population, then some of the population is exactly equal to the median. For example, if $a < b < c$, then the median of the list $\{a, b, c\}$ is b, and, if $a < b < c < d$, then the median of the list $\{a, b, c, d\}$ is the mean of b and c; i.e., it is $(b + c)/2$.

The median can be used as a measure of location when a distribution is skewed, when end-values are not known, or when one requires reduced importance to be attached to outliers, e.g., because they may be measurement errors. In terms of notation, some authors represent the median of a variable x either as $\tilde{x}$ or as $\mu_{1/2}$, sometimes also M. There is no widely accepted standard notation for the median, so the use of these or other symbols for the median needs to be explicitly defined when they are introduced.

The median is the 2nd quartile, 5th decile, and 50th percentile.

Measures of Location and Dispersion

The median is one of a number of ways of summarising the typical values associated with members of a statistical population; thus, it is a possible location parameter. When the median is used as a location parameter in descriptive statistics, there are several choices for a measure of variability: the range, the interquartile range, the mean absolute deviation, and the median absolute deviation. Since the median is the same as the *second quartile*, its calculation is illustrated in the article on quartiles.

For practical purposes, different measures of location and dispersion are often compared on the basis of how well the corresponding population values can be estimated from a sample of data. The median, estimated using the sample median, has good properties in this regard. While it is not usually optimal if a given population distribution is assumed, its properties are always reasonably good.

For example, a comparison of the efficiency of candidate estimators shows that the sample mean is more statistically efficient than the sample median when data are uncontaminated by data from heavy-tailed distributions or from mixtures of distributions, but less efficient otherwise, and that the efficiency of the sample median is higher than that for a wide range of distributions. More specifically, the median has a 64% efficiency compared to the minimum-variance mean (for large normal samples), which is to say the variance of the median will be ~50% greater than the variance of the mean.

Medians of Probability Distributions

For any probability distribution on the real line R with cumulative distribution function *F*, regardless of whether it is any kind of continuous probability distribution, in particular an absolutely continuous distribution (which has a probability density function), or a discrete probability distribution, a median is by definition any real number *m* that satisfies the inequalities

$$P(X \leq m) \geq \frac{1}{2} \text{ and } P(X \geq m) \geq \frac{1}{2}$$

or, equivalently, the inequalities

$$\int_{-\infty}^{m} dF(x) \geq \frac{1}{2} \text{ and } \int_{m}^{\infty} dF(x) \geq \frac{1}{2}$$

in which a Lebesgue–Stieltjes integral is used. For an absolutely continuous probability distribution with probability density function f, the median satisfies

$$P(X \le m) = P(X \ge m) = \int_{-\infty}^{m} f(x)dx = \frac{1}{2}.$$

Any probability distribution on R has at least one median, but there may be more than one median. Where exactly one median exists, statisticians speak of "the median" correctly; even when the median is not unique, some statisticians speak of "the median" informally.

Medians of Particular Distributions

The medians of certain types of distributions can be easily calculated from their parameters:

- The median of a symmetric distribution with mean μ is μ.
  - — The median of a normal distribution with mean μ and variance σ^2 is μ. In fact, for a normal distribution, mean = median = mode.
  - — The median of a uniform distribution in the interval $[a, b]$ is $(a + b)/2$, which is also the mean.
- The median of a Cauchy distribution with location parameter x_0 and scale parameter y is x_0, the location parameter.
- The median of an exponential distribution with rate parameter is the natural logarithm of 2 divided by the rate parameter: $\lambda^{-1} \ln 2$.
- The median of a Weibull distribution with shape parameter k and scale parameter λ is $\lambda(\ln 2)^{1/k}$.

Medians in Descriptive Statistics

The median is used primarily for skewed distributions, which it summarizes differently than the arithmetic mean. Consider the multiset {1, 2, 2, 2, 3, 14}. The median is 2 in this case, as is the mode, and it might be seen as a better indication of central tendency than the arithmetic mean of 4. Calculation of medians is a popular technique in summary statistics and summarizing statistical data, since it is simple to understand and easy to calculate, while also giving a measure that is more robust in the presence of outlier values than is the mean.

Medians for Populations

An Optimality Property: The *mean absolute error* of a real variable c with respect to the random variable X is

$$E(|X - c|) .$$

Provided that the probability distribution of X is such that the above expectation exists, then m is a median of X if and only if m is a minimizer of the mean absolute error with respect to X. In particular, m is a sample median if and only if m minimizes the arithmetic mean of the absolute deviations.

Unimodal Distributions

It can be shown for a unimodal distribution that the mean and the median lie within $(3/5)^{1/2}$ times the standard deviation of each other. In symbols

$$\frac{|\tilde{x} - \bar{x}|}{\sigma} \leq (3/5)^{1/2}$$

where |.| is the absolute value. A similar relation holds between the mode and the median. In this case the mode and the median lie within $3^{1/2}$ times the standard deviation of each other.

$$\frac{|\tilde{x} - \text{mode}|}{\sigma} \leq 3^{1/2}$$

An Inequality Relating Means and Medians

If the distribution has finite variance, then the distance between the median and the mean is bounded by one standard deviation.

This bound was proved by Mallows, who used Jensen's inequality twice, as follows. We have

$$\begin{aligned} |\mu - m| = |\mathrm{E}(X - m)| &\leq \mathrm{E}(|X - m|) \\ &\leq \mathrm{E}(|X - \mu|) \\ &\leq \sqrt{\mathrm{E}((X - \mu)^2)} = \sigma. \end{aligned}$$

The first and third inequalities come from Jensen's inequality applied to the absolute-value function and the square function, which are each convex. The second inequality comes from the fact that a median minimizes the absolute deviation function

$$a \mapsto \mathrm{E}(|X - a|).$$

This proof can easily be generalized to obtain a multivariate version of the inequality, as follows:

$$\begin{aligned} \|\mu - m\| = \|E(X - m)\| &\leq E \ X - m \\ &\leq E(\|X - \mu\|) \\ &\leq \sqrt{E(\ X - \mu\ ^2)} = \sqrt{trace(\mathrm{var}(X))} \end{aligned}$$

where m is a spatial median, that is, a minimizer of the function $a \mapsto E(\|X-a\|)$. The spatial median is unique when the data-set's dimension is two or more. An alternative proof uses the one-sided Chebyshev inequality; it appears in an inequality on location and scale parameters.

Jensen's Inequality for Medians

Jensen's inequality states that for any random variable x with a ûnite expectation $E(X)$ and for any convex function f then

$$f(E(x)) \leq E(f(x))$$

It has been shown that if x is a real variable with a unique median m and f is a C function then

$$f(m) \leq \text{Median}(f(x))$$

AC function is a real valued function, defined on the set of real numbers R, with the property that for any real t

$$f^{-1}((-\infty,t]) = \{x \in R \mid f(x) \leq t\}$$

is a closed interval, a singleton or an empty set.

Medians for Samples

The Sample Median

Efficient Computation of the Sample Median: Even though sorting n items requires $O(n \log n)$ operations, selection algorithms can compute the kth-smallest of n items (e.g., the median) with only $O(n)$ operations.

Easy Explanation of the Sample Median: In individual series (if number of observation is very low) first one must arrange all the observations in ascending order. Then count(n) total number of observation in given data.

If n is odd then Median (M) = value of ((n + 1)/2)th item term.

If n is even then Median (M) = value of [((n)/2)th item term + ((n)/2 + 1)th item term]/2

For an odd number of values: As an Example, we will Calculate the Sample Median for the following set of Observations: 1, 5, 2, 8, 7.

Start by Sorting the Values: 1, 2, 5, 7, 8.

In this case, the median is 5 since it is the middle observation in the ordered list.

The median is the $((n + 1)/2)$th item, where n is the number of values. For example, for the list {1, 2, 5, 7, 8}, we have $n = 5$, so the median is the ((5 + 1)/2)th item.

median = (6/2)th item

median = 3rd item

median = 5

For an even number of values: As an Example, we will Calculate the Sample Median for the following set of Observations: 1, 6, 2, 8, 7, 2.

Start by Sorting the Values: 1, 2, 2, 6, 7, 8.

In this case, the arithmetic mean of the two middlemost terms is (2 + 6)/2 = 4. Therefore, the median is 4 since it is the arithmetic mean of the middle observations in the ordered list.

We also use this formula MEDIAN = $\{(n + 1)/2\}$th item .n = number of values

As above example 1, 2, 2, 6, 7, 8 $n = 6$ Median = {(6 + 1)/2}th item = 3.5th item. In this case, the median is average of the 3rd number and the next one (the fourth number). The median is (2 + 6)/2 which is 4.

Variance

The distribution of both the sample mean and the sample median were determined by Laplace.

The distribution of the sample median from a population with a density function $f(x)$ is asymptotically normal with mean m and variance is

$$\frac{1}{4nf(m)^2}$$

where m is the median value of distribution and n is the sample size. In practice this may be difficult to estimate as the density function is usually unknown.

These results have also been extended. It is now known that for the pth quartile that the distribution of the sample p^{th} quartile is distributed normally around the pth quartile with variance equal to

$$\frac{p(1-p)}{nf(x_p)^2}$$

where $f(x_p)$ is the value of the distribution at the pth quartile.

Estimation of variance from sample data: The value of $(2f(\ x\ )\ )^{-2}$ – the asymptotic value of $n^{-1/2}(\nu - m)$ where ν is the population median – has been studied by several authors. The standard 'delete one' jackknife method produces inconsistent results. An alternative – the 'delete k' method – where k grows with the sample size has been shown to be asymptotically consistent. This method may be computationally expensive for large data sets. A bootstrap estimate is known to be consistent but converges very slowly (order of $n^{-1/4}$) Other methods have been proposed but their behaviour may differ between large and small samples.

Efficiency

The efficiency of the sample median, measured as the ratio of the variance of the mean to the variance of the median, depends on the sample size and on the underlying population distribution. For a sample of size $N = 2n + 1$ from the normal distribution, the ratio is

$$\frac{4n}{\pi(2n+1)}$$

For large samples (as n tends to infinity) this ratio tends to $\frac{2}{\pi}$.

Other Estimators

For univariate distributions that are *symmetric* about one median, the Hodges–Lehmann estimator is a robust and highly efficient estimator of the population median.

If data are represented by a statistical model specifying a particular family of probability distributions, then estimates of the median can be obtained by fitting that family of probability distributions to the data and calculating the theoretical median of the fitted distribution. Pareto interpolation is an applicaton of this when the population is assumed to have a Pareto distribution.

Coefficient of Dispersion

The coefficient of dispersion (CD) is defined as the ratio of the average absolute deviation from the median to the median of the data. It is a statistical measure used by the states of Iowa, New York and South Dakota in estimating dues taxes. In symbols

$$CD = \frac{1}{n}\frac{\sum |m - x|}{m}$$

where n is the sample size, m is the sample median and x is a variate.

The sum is taken over the whole sample. Confidence intervals for a two sample test where the sample sizes are large have been derived by Bonett and Seier This test assumes that both sames have the same median but differ in the dispersion around it. The confidence interval (CI) is bounded inferiorly by

$$\exp[\log(\frac{t_a}{t_b}) - z_\alpha (var[\log(\frac{t_a}{t_b})])^{0.5}]$$

where t_j is the mean absolute deviation of the jth sample, $var()$ is the variance and z_α is the value from the normal distribution for the chosen value of α: for $\alpha = 0.05$, $z_\alpha = 1.96$. The following formulae are used in the derivation of these confidence intervals

$$var[\log(t_a)] = \frac{(\frac{s_a^2}{t_a^2} + (\frac{x_a - \overline{x}}{t_a})^2 - 1)}{n}$$

$$var[\log(t_a / t_b)] = var[\log(t_a)] + var[\log(t_b)] - 2r(var[\log(t_a)]var[\log(t_b)])^{0.5}$$

where r is the Pearson correlation coefficient between the squared deviation scores

$$d_{ia} = | x_{ia} - \overline{x}_a | \text{ and } d_{ib} = | x_{ib} - \overline{x}_b |$$

a and b here are constants equal to 1 and 2, x is a variate and s is the standard deviation of the sample.

Multivariate Median

Previously, this article discussed the concept of a univariate median for a one-dimensional object (population, sample). When the dimension is two or higher, there are multiple concepts that extend the definition of the univariate median; each such multivariate median agrees with the univariate median when the dimension is exactly one. In higher dimensions, however, there are several multivariate medians.

Marginal Median

The marginal median is defined for vectors defined with respect to a fixed set of coordinates. A marginal median is defined to be the vector whose components are univariate medians. The marginal median is easy to compute, and its properties were studied by Puri and Sen.

Spatial Median (L1 Median)

In a normed vector space of dimension two or greater, the "spatial median" minimizes the expected distance

$$a \mapsto \mathrm{E}(\|X - a\|),$$

where X and a are vectors, if this expectation has a finite minimum; another definition is better suited for general probability-distributions. The spatial median is unique when the data-set's dimension is two or more. It is a robust and highly efficient estimator of the population spatial-median (also called the "L1 median").

Other Multivariate Medians

An alternative to the spatial median is defined in a similar way, but based on a different loss function, and is called the Geometric median. The centrepoint is another generalization to higher dimensions that does not relate to a particular metric.

Other Median-related Concepts

Pseudo-median

For univariate distributions that are *symmetric* about one median, the Hodges–Lehmann estimator is a robust and highly efficient estimator of the population median; for non-symmetric distributions, the Hodges–Lehmann estimator is a robust and highly efficient estimator of the population *pseudo-median*, which is the median of a symmetrized distribution and which is close to the population median. The Hodges–Lehmann estimator has been generalized to multivariate distributions.

Variants of Regression

The Theil–Sen estimator is a method for robustlinear regression based on finding medians of slopes.

Median Filter

In the context of image processing of monochromeraster images there is a type of noise, known as the salt and pepper noise, when each pixel independently becomes black (with some small probability) or white (with some small probability), and is unchanged otherwise (with the probability close to 1). An image constructed of median values of neighbourhoods (like 3×3 square) can effectively reduce noise in this case.

Cluster Analysis

In cluster analysis, the k-medians clustering algorithm provides a way of defining clusters, in which the criterion of maximising the distance between cluster-means that is used in k-means clustering, is replaced by maximising the distance between cluster-medians.

Median-unbiased Estimators

Any *mean*-unbiased estimator minimizes the risk (expected loss) with respect to the squared-error loss function, as observed by Gauss. A *median*-unbiased estimator minimizes the risk with respect to the absolute-deviation loss function, as observed by Laplace. Other loss functions are used in statistical theory, particularly in robust statistics.

The theory of median-unbiased estimators was revived by George W. Brown in 1947:

An estimate of a one-dimensional parameter θ will be said to be median-unbiased if, for fixed θ, the median of the distribution of the estimate is at the value θ; i.e., the estimate underestimates just as often as it overestimates. This requirement seems for most purposes to accomplish as much as the mean-unbiased requirement and has the additional property that it is invariant under one-to-one transformation. Further properties of median-unbiased estimators have been reported. In particular, median-unbiased estimators exist in cases where mean-unbiased and maximum-likelihood estimators do not exist. Median-unbiased estimators are invariant under one-to-one transformations.

History

The idea of the median originated in Edward Wright's book on navigation (*Certaine Errors in Navigation*) in 1599 in a section concerning the determination of location with a compass. Wright felt that this value was the most likely to be the correct value in a series of observations.

In 1757, Roger Joseph Boscovich developed a regression method based on the *L*1 norm and therefore implicitly on the median. The distribution of both the sample mean and the sample median were determined by Laplace in the early 1800s. Antoine AugustinCournot in 1843 was the first to use the term *median* (*valeurmédiane*) for the value that divides a probability distribution into two equal halves. Gustav Theodor Fechner used the median (*Centralwerth*) in sociological and psychological phenomena. It had earlier been used only in astronomy and related fields. Gustav Fechner popularized the median into the formal analysis of data, although it had been used previously by Laplace.

Francis Galton used the English term *median* in 1881, having earlier used the terms *middle-most value* in 1869 and the *medium* in 1880.

Chapter

4

Complex Exponents with Positive Real Bases

Imaginary Exponents with Base e

The geometric interpretation of the operations on complex numbers and the definition of the exponential function is the clue to understanding e^{ix} for real x. Consider the right triangle $(0, 1, 1 + ix/n)$. For big values of n the triangle is almost a circular sector with a small central angle equal to x/nradians.

The triangles $(0, (1 + ix/n)^k, (1 + ix/n)^{k+1})$ are mutually similar for all values of k. So for large values of n the limiting point of $(1 + ix/n)^n$ is the point on the unit circle whose angle from the positive real axis is x radians. The polar coordinates of this point are $(r, \theta) = (1, x)$, and the cartesian coordinates are $(\cos x, \sin x)$. So $e^{ix} = \cos x + i \sin x$, and this is Euler's formula, connecting algebra to trigonometry by means of complex numbers.

The solutions to the equation $e^z = 1$ are the integer multiples of $2\pi i$:

$$\{z : e^z = 1\} = \{2k\pi i : k \in \mathbb{Z}\}$$

More generally, if $e^v = w$, then every solution to $e^z = w$ can be obtained by adding an integer multiple of $2\pi i$ to v:

$$\{z : e^z = w\} = \{v + 2k\pi i : k \in \mathbb{Z}\}$$

Thus the complex exponential function is a periodic function with period $2\pi i$

More Simply: $e^{i\pi} = -1$; $e^{x+iy} = e^x(\cos y + i \sin y)$.

Trigonometric Functions

It follows from Euler's formula stated above that the trigonometric functions cosine and sine are

$$\cos(z) = \frac{e^{iz} + e^{-iz}}{2}; \qquad \sin(z) = \frac{e^{iz} - e^{-iz}}{2i}$$

Historically, cosine and sine were defined geometrically before the invention of complex numbers. The above formula reduces the complicated formulas for trigonometric functions of a sum into the simple exponentiation formula

$$e^{i(x+y)} = e^{ix} \cdot e^{iy}$$

Using exponentiation with complex exponents may reduce problems in trigonometry to algebra.

Complex Exponents with Base e

The power $z = e^{x+iy}$ can be computed as $e^x \cdot e^{iy}$. The real factor e^x is the absolute value of z and the complex factor e^{iy} identifies the direction of z.

Complex Exponents with Positive Real Bases

If b is a positive real number, and z is any complex number, the power b^z is defined as $e^{z \ln(b)}$, where $x = \ln(b)$ is the unique real solution to the equation $e^x = b$. So the same method working for real exponents also works for complex exponents.

For example:

$2^i = e^{i \ln(2)} = \cos(\ln(2)) + i \sin(\ln(2)) \approx 0.76924 + 0.63896i$

$e^i \approx 0.54030 + 0.84147i$

$10^i \approx -0.66820 + 0.74398i$

$(e^{2\pi})^i \approx 535.49^i \approx 1$

The identity $(b^z)^u = b^{zu}$ is not generally valid for complex powers. A simple counterexample is given by:

$$(e^{2\pi i})^i = 1^i = 1 \neq e^{-2\pi} = e^{2\pi i \cdot i}$$

The identity is, however, valid when z is a real number, and also when u is an integer.

Powers of Complex Numbers

Integer powers of nonzero complex numbers are defined by repeated multiplication or division as above. If i is the imaginary unit and n is an integer, then i^n equals 1, i, –1, or “i, according to whether the integer n is congruent to 0, 1, 2, or 3 modulo 4. Because of this, the powers of i are useful for expressing sequences of period 4.

Complex powers of positive reals are defined via e^x as in section Complex powers of positive real numbers above. These are continuous functions.

Trying to extend these functions to the general case of noninteger powers of complex numbers that are not positive reals leads to difficulties. Either we define discontinuous functions or multivalued functions. Neither of these options is entirely satisfactory.

The rational power of a complex number must be the solution to an algebraic equation. Therefore it always has a finite number of possible values. For example, $w = z^{1/2}$ must be a solution to the equation $w^2 = z$. But if w is a solution, then so is $-w$, because $(-1)^2 = 1$. A unique but somewhat arbitrary solution called the principal value can be chosen using a general rule which also applies for nonrational powers.

Complex powers and logarithms are more naturally handled as single valued functions on a Riemann surface. Single valued versions are defined by choosing a sheet. The value has a discontinuity along a branch cut. Choosing one out of many solutions as the principal value leaves us with functions that are not continuous, and the usual rules for manipulating powers can lead us astray.

Any nonrational power of a complex number has an infinite number of possible values because of the multi-valued nature of the complex logarithm. The principal value is a single value chosen from these by a rule which, amongst its other properties, ensures powers of complex numbers with a positive real part and zero imaginary part give the same value as for the corresponding real numbers.

Exponentiating a real number to a complex power is formally a different operation from that for the corresponding complex number. However in the common case of a positive real number the principal value is the same.

The powers of negative real numbers are not always defined and are discontinuous even where defined. When dealing with complex numbers the complex number operation is normally used instead.

Complex Power of a Complex Number

For complex numbers w and z with $w \neq 0$, the notation w^z is ambiguous in the same sense that log w is. To obtain a value of w^z, first choose a logarithm of w; call it log w. Such a choice may be the principal value Log w (the default, if no other specification is given), or perhaps a value given by some other branch of log w fixed in advance. Then, using the complex exponential function one defines

$$w^z = e^{z \log w}$$

because this agrees with the earlier definition in the case where w is a positive real number and the (real) principal value of log w is used.

If z is an integer, then the value of w^z is independent of the choice of log w, and it agrees with the earlier definition of exponentation with an integer exponent.

If z is a rational numberm/n in lowest terms with $z > 0$, then the infinitely many choices of log w yield only n different values for w^z; these values are the n complex solutions s to the equation $s^n = w^m$.

If z is an irrational number, then the infinitely many choices of log w lead to infinitely many distinct values for w^z.

The computation of complex powers is facilitated by converting the base w to polar form, as described in detail below. A similar construction is employed in quaternions.

Complex Roots of Unity

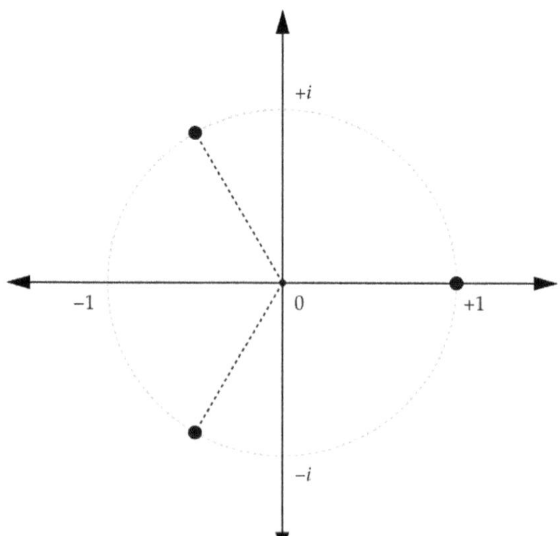

***Figure:** The three 3rd roots of 1*

A complex number w such that $w^n = 1$ for a positive integer n is an nth root of unity. Geometrically, the nth roots of unity lie on the unit circle of the complex plane at the vertices of a regular n-gon with one vertex on the real number 1.

If $w^n = 1$ but $w^k \neq 1$ for all natural numbers k such that $0 < k < n$, then w is called a primitive nth root of unity. The negative unit "1 is the only primitive square root of unity. The imaginary unit$i$ is one of the two primitive 4-th roots of unity; the other one is $-i$.

The number $e^{2\pi i (1/n)}$ is the primitive nth root of unity with the smallest positive complex argument. (It is sometimes called the principal nth root of unity, although this terminology is not universal and should not be confused with the principal value of $\sqrt[n]{1}$, which is 1.)

The other nth roots of unity are given by

$$\left(e^{\frac{2}{n}\pi i}\right)^k = e^{\frac{2}{n}\pi ik}$$

for $2 \leq k \leq n$.

Roots of Arbitrary Complex Numbers

Although there are infinitely many possible values for a general complex logarithm, there are only a finite number of values for the power w^q in the important special case where $q = 1/n$ and n is a positive integer. These are the nth roots of w; they are solutions of the equation $z^n = w$. As with real roots, a second root is also called a square root and a third root is also called a cube root.

It is conventional in mathematics to define $w^{1/n}$ as the principal value of the root. If w is a positive real number, it is also conventional to select a positive real number as the principal value of the root $w^{1/n}$. For general complex numbers, the nth root with the smallest argument is often selected as the principal value of the nth root operation, as with principal values of roots of unity.

The set of nth roots of a complex number w is obtained by multiplying the principal value $w^{1/n}$ by each of the nth roots of unity. For example, the fourth roots of 16 are 2, –2, $2i$, and $-2i$, because the principal value of the fourth root of 16 is 2 and the fourth roots of unity are 1, –1, i, and $-i$.

Computing Complex Powers

It is often easier to compute complex powers by writing the number to be exponentiated in polar form. Every complex number z

can be written in the polar form

$$z = re^{i\theta} = e^{\ln(r)+i\theta}$$

where r is a nonnegative real number and θ is the (real) argument of z. The polar form has a simple geometric interpretation: if a complex number $u + iv$ is thought of as representing a point (u, v) in the complex plane using Cartesian coordinates, then (r, θ) is the same point in polar coordinates. That is, r is the "radius" $r^2 = u^2 + v^2$ and θ is the "angle" $\theta = \text{atan2}(v, u)$. The polar angle θ is ambiguous since any multiple of 2π could be added to θ without changing the location of the point. Each choice of θ gives in general a different possible value of the power. A branch cut can be used to choose a specific value. The principal value (the most common branch cut), corresponds to θ chosen in the interval $(-\pi, \pi]$. For complex numbers with a positive real part and zero imaginary part using the principal value gives the same result as using the corresponding real number.

In order to compute the complex power w^z, write w in polar form:

$$w = re^{i\theta}$$

Then

$$\log w = \log r + i\theta$$

and thus

$$w^z = e^{z \log w} = e^{z(\log r + i\theta)}$$

If z is decomposed as $c + di$, then the formula for w^z can be written more explicitly as

$$\left(r^c e^{-d\theta}\right)e^{i(d \log r + c\theta)} = \left(r^c e^{-d\theta}\right)\left[\cos(d \log r + c\theta) + i \sin(d \log r + c\theta)\right]$$

This final formula allows complex powers to be computed easily from decompositions of the base into polar form and the exponent into Cartesian form. It is shown here both in polar form and in Cartesian form (via Euler's identity). The following examples use the principal value, the branch cut which causes θ to be in the interval $(-\pi, \pi]$. To compute i^i, write i in polar and Cartesian forms:

$$i = 1 \cdot e^{\frac{1}{2}i\pi}$$
$$i = 0 + 1i$$

Then the formula above, with $r = 1$, $\theta = \pi/_2$, $c = 0$, and $d = 1$, yields:

$$i^i = \left(1^0 e^{-\frac{1}{2}\pi}\right)e^{i\left[1 \cdot \log 1 + 0 \cdot \frac{1}{2}\pi\right]} = e^{-\frac{1}{2}\pi} \approx 0.2079$$

Similarly, to find $(-2)^{3+4i}$, compute the polar form of –2,

$$-2 = 2e^{i\pi}$$

and use the formula above to compute

$$(-2)^{3+4i} = \left(2^3 e^{-4\pi}\right) e^{i[4\log(2)+3\pi]} \approx (2.602 - 1.006i)\cdot 10^{-5}$$

The value of a complex power depends on the branch used. For example, if the polar form $i = 1e^{i(5\pi/2)}$ is used to compute i^i, the power is found to be $e^{-5\pi/2}$; the principal value of i^i, computed above, is $e^{-\pi/2}$. The set of all possible values for i^i is given by:

$$\begin{aligned} i &= 1\cdot e^{\frac{1}{2}i\pi + i2\pi k} \,\Big|\, k \in \mathbb{Z} \\ i^i &= e^{i\left(\frac{1}{2}i\pi + i2\pi k\right)} \\ &= e^{-\left(\frac{1}{2}\pi + 2\pi k\right)} \end{aligned}$$

So there is an infinity of values which are possible candidates for the value of i^i, one for each integer k. All of them have a zero imaginary part so one can say i^i has an infinity of valid real values.

Failure of Power and Logarithm Identities

Some identities for powers and logarithms for positive real numbers will fail for complex numbers, no matter how complex powers and complex logarithms are defined *as single-valued functions*. For example:

- The identity $\log(b^x) = x \cdot \log\ b$ holds whenever b is a positive real number and x is a real number. But for the principal branch of the complex logarithm one has

  $$i\pi = \log(-1) = \log\left[(-i)^2\right] \neq 2\log(-i) = 2\left(-\frac{i\pi}{2}\right) = -i\pi$$

  Regardless of which branch of the logarithm is used, a similar failure of the identity will exist. The best that can be said (if only using this result) is that:

  $$\log(w^z) \equiv z\cdot\log(w) \pmod{2\pi i}$$

  This identity does not hold even when considering log as a multivalued function. The possible values of $\log(w^z)$ contain those of $z \cdot \log w$ as a subset. Using $\mathrm{Log}(w)$ for the principal value of $\log(w)$ and m, n as any integers the possible values of both sides are:

$$\{\log(w^z)\} = \{z{\cdot}\mathrm{Log}(w) + z{\cdot}2\pi in + 2\pi im\}$$

$$\{z{\cdot}\log(w)\} = \{z{\cdot}\mathrm{Log}(w) + z{\cdot}2\pi in\}$$

- The identities $(bc)^x = b^x c^x$ and $(b/c)^x = b^x/c^x$ are valid when b and c are positive real numbers and x is a real number. But a calculation using principal branches shows that

  $$1 = (-1\times-1)^{\frac{1}{2}} \neq (-1)^{\frac{1}{2}}(-1)^{\frac{1}{2}} = -1$$

  and

  $$i = (-1)^{\frac{1}{2}} = \left(\frac{1}{-1}\right)^{\frac{1}{2}} \neq \frac{1^{\frac{1}{2}}}{(-1)^{\frac{1}{2}}} = \frac{1}{i} = -i$$

  On the other hand, when x is an integer, the identities are valid for all nonzero complex numbers.

  If exponentiation is considered as a multivalued function then the possible values of $(-1\times-1)^{1/2}$ are $\{1, -1\}$. The identity holds but saying $\{1\} = \{(-1\times-1)^{1/2}\}$ is wrong.

- The identity $(e^x)^y = e^{xy}$ holds for real numbers x and y, but assuming its truth for complex numbers leads to the following paradox, discovered in 1827 by Clausen:

  For any integer n, we have:

  $$e^{1+2\pi in} = e^1 e^{2\pi in} = e{\cdot}1 = e$$

  $$\left(e^{1+2\pi in}\right)^{1+2\pi in} = e$$

  $$e^{1+4\pi in-4\pi^2n^2} = e$$

  $$e^1 e^{4\pi in} e^{-4\pi^2n^2} = e$$

  $$e^{-4\pi^2n^2} = 1$$

but this is false when the integer n is nonzero.

There are a number of problems in the reasoning:

The major error is that changing the order of exponentiation in going from line two to three changes what the principal value chosen will be.

From the multi-valued point of view, the first error occurs even sooner. Implicit in the first line is that e is a real number, whereas the result of $e^{1+2\pi in}$ is a complex number better represented as $e+0i$. Substituting the complex number for the real on the second line makes the power have multiple possible values. Changing the order of exponentiation from lines two to three also affects how many possible values the result can have. $(e^z)^w \neq e^{zw}$, but rather $(e^z)^w = e^{(z+2\pi in)w}$ multivalued over integers n.

Zero to the Power of Zero

For Discrete Exponents

In most settings not involving continuity in the exponent, interpreting 0^0 as 1 simplifies formulas and eliminates the need for special cases in theorems. For example:

- Regarding b^0 as an empty product assigns it the value 1, even when $b = 0$.
- The combinatorial interpretation of 0^0 is the number of empty tuples of elements from the empty set. There is exactly one empty tuple.
- Equivalently, the set-theoretic interpretation of 0^0 is the number of functions from the empty set to the empty set. There is exactly one such function, the empty function.
- The notation $\sum a_n x^n$ for polynomials and power series rely on defining $0^0 = 1$. Identities like $\frac{1}{1-x} = \sum_{n=0}^{\infty} x^n$ and $e^x = \sum_{n=0}^{\infty} \frac{x^n}{n!}$ and the binomial theorem $(1+x)^n = \sum_{k=0}^{n} \binom{n}{k} x^k$ are not valid for $x = 0$ unless $0^0 = 1$.
- In differential calculus, the power rule $\frac{d}{dx} x^n = nx^{n-1}$ is not valid for $n = 1$ at $x = 0$ unless $0^0 = 1$.

In Analysis

On the other hand, when 0^0 arises when trying to determine a limit of the form $\lim_{x \to 0} f(x)^{g(x)}$, it must be handled as an indeterminate form.

- Limits involving algebraic operations can often be evaluated by replacing subexpressions by their limits; if the resulting expression does not determine the original limit, the expression is known as an indeterminate form. In fact, when $f(t)$ and $g(t)$ are real-valued functions both approaching 0 (as t approaches a real number or $\pm\infty$), with $f(t) > 0$, the function $f(t)^{g(t)}$ need not approach 1; depending on f and g, the limit of $f(t)^{g(t)}$ can be any nonnegative real number or $+\infty$, or it can be undefined. For example, the functions below are of the form $f(t)^{g(t)}$ with $f(t),g(t) \to 0$ as $t \to 0^+$, but the limits are different:

$$\lim_{t\to 0^+} t^t = 1,\quad \lim_{t\to 0^+}\left(e^{-\frac{1}{t^2}}\right)^t = 0,\quad \lim_{t\to 0^+}\left(e^{-\frac{1}{t^2}}\right)^{-t} = +\infty,\quad \lim_{t\to 0^+}\left(e^{-\frac{1}{t}}\right)^{at} = e^{-a}.$$

So 0^0 is an indeterminate form. This behaviour shows that the two-variable function x^y, though continuous on the set $\{(x,y): x > 0\}$, cannot be extended to a continuous function on any set containing $(0,0)$, no matter how 0^0 is defined. However, under certain conditions, such as when f and g are both analytic functions and f is nonnegative, the limit approaching from the right is always 1.

- In the complex domain, the function z^w is defined for nonzero z by choosing a branch of $\log z$ and setting $z^w := e^{w \log z}$, but there is no branch of $\log z$ defined at $z = 0$, let alone in a neighbourhood of 0.

History of Differing Points of View

Different authors interpret the situation above in different ways:

- Some argue that the best value for 0^0 depends on context, and hence that defining it once and for all is problematic. According to Benson (1999), "The choice whether to define 0^0 is based on convenience, not on correctness."
- Others argue that 0^0 is 1. According to p. 408 of Knuth (1992), it "*has* to be 1", although he goes on to say that "Cauchy had good reason to consider 0^0 as an undefined *limiting form*" and that "in this much stronger sense, the value of 0^0 is less defined than, say, the value of $0 + 0$" (emphases in original).

The debate has been going on at least since the early 19th century. At that time, most mathematicians agreed that $0^0 = 1$, until in 1821 Cauchy listed 0^0 along with expressions like $^0/_0$ in a table of undefined forms. In the 1830s Libri published an unconvincing argument for $0^0 = 1$, and Möbius sided with him, erroneously claiming that $\lim_{t\to 0^+} f(t)^{g(t)} = 1$ whenever $\lim_{t\to 0^+} f(t) = \lim_{t\to 0^+} g(t) = 0$. A commentator who signed his name simply as "S" provided the counterexample of $(e^{-1/t})^t$, and this quieted the debate for some time, with the apparent conclusion of this episode being that 0^0 should be undefined. More details can be found in Knuth (1992).

Treatment on computers

IEEE Floating Point Standard

The IEEE 754-2008 floating point standard is used in the design of most floating point libraries. It recommends a number of different functions for computing a power:

- pow treats 0^0 as 1. This is the oldest defined version. If the power is an exact integer the result is the same as for pown, otherwise the result is as for powr (except for some exceptional cases).
- pown treats 0^0 as 1. The power must be an exact integer. The value is defined for negative bases; e.g., pown(–3,5) is "243.
- powr treats 0^0 as NaN (Not-a-Number – undefined). The value is also NaN for cases like powr(–3,2) where the base is less than zero. The value is defined by $e^{\text{power'}\times\log(\text{base})}$.

Programming Languages

Most programming language with a power function are implemented using the IEEE pow function and therefore evaluate 0^0 as 1. The later C and C++ standards describe this as the normative behaviour. The Java standard mandates this behaviour. The .NET Framework method System. Math.Pow also treats 0^0 as 1.

Mathematics Software

- Sage simplifies b^0 to 1, even if no constraints are placed on b. It does not simplify 0^x, and it takes 0^0 to be 1.
- Maple simplifies b^0 to 1 and 0^x to 0, even if no constraints are placed on b (the latter simplification is only valid for $x > 0$), and evaluates 0^0 to 1.
- Macsyma also simplifies b^0 to 1 and 0^x to 0, even if no constraints are placed on b and x, but issues an error for 0^0.
- Mathematica and Wolfram Alpha simplify b^0 into 1, even if no constraints are placed on b. While Mathematica does not simplify 0^x, Wolfram Alpha returns two results, 0 and indeterminate. Both Mathematica and Wolfram Alpha take 0^0 to be an indeterminate form.
- Qalculate! evaluates 0^0 as 1, however it displays a warning notifying the user that *"0^0 might be considered undefined"*.

Limits of Powers

The section zero to the power of zero gives a number of examples of limits which are of the indeterminate form 0^0. The limits in these examples exist, but have different values, showing that the two-variable function x^y has no limit at the point (0,0). One may ask at what points this function does have a limit.

More precisely, consider the function $f(x,y) = x^y$ defined on $D = \{(x,y) \in R^2 : x > 0\}$. Then D can be viewed as a subset of R^2 (that is, the set of all pairs (x,y) with x,y belonging to the extended real number lineR = $[-\infty, +\infty]$, endowed with the product topology), which will contain the points at which the function f has a limit.

In fact, f has a limit at all accumulation points of D, except for (0,0), $(+\infty,0)$, $(1,+\infty)$ and $(1,-\infty)$. Accordingly, this allows one to define the powers x^y by continuity whenever $0 \le x \le +\infty$, $-\infty \le y \le +\infty$, except for 0^0, $(+\infty)^0$, $1^{+\infty}$ and $1^{-\infty}$, which remain indeterminate forms.

Under this definition by continuity, we obtain:

- $x^{+\infty} = +\infty$ and $x^{-\infty} = 0$, when $1 < x \le +\infty$.
- $x^{+\infty} = 0$ and $x^{-\infty} = +\infty$, when $0 \le x < 1$.
- $0^y = 0$ and $(+\infty)^y = +\infty$, when $0 < y \le +\infty$.
- $0^y = +\infty$ and $(+\infty)^y = 0$, when $-\infty \le y < 0$.

These powers are obtained by taking limits of x^y for *positive* values of x. This method does not permit a definition of x^y when $x < 0$, since pairs (x,y) with $x < 0$ are not accumulation points of D.

On the other hand, when n is an integer, the power x^n is already meaningful for all values of x, including negative ones. This may make the definition $0^n = +\infty$ obtained above for negative n problematic when n is odd, since in this case $x^n \to +\infty$ as x tends to 0 through positive values, but not negative ones.

Efficient Computation of Integer Powers

The simplest method of computing b^n requires $n-1$ multiplication operations, but it can be computed more efficiently than that, as illustrated by the following example. To compute 2^{100}, note that 100 = 64 + 32 + 4. Compute the following in order:

1. $2^2 = 4$
2. $(2^2)^2 = 2^4 = 16$
3. $(2^4)^2 = 2^8 = 256$
4. $(2^8)^2 = 2^{16} = 65{,}536$

5. $(2^{16})^2 = 2^{32} = 4{,}294{,}967{,}296$
6. $(2^{32})^2 = 2^{64} = 18{,}446{,}744{,}073{,}709{,}551{,}616$
7. $2^{64}\ 2^{32}\ 2^4 = 2^{100}$ = 1, 267, 650, 600, 228, 229, 401, 496, 703, 205,376.

This series of steps only requires 8 multiplication operations instead of 99 (since the last product above takes 2 multiplications).

In general, the number of multiplication operations required to compute b^n can be reduced to $\Theta(\log n)$ by using exponentiation by squaring or (more generally) addition-chain exponentiation. Finding the *minimal* sequence of multiplications (the minimal-length addition chain for the exponent) for b^n is a difficult problem for which no efficient algorithms are currently known (see Subset sum problem), but many reasonably efficient heuristic algorithms are available.

Exponential Notation for Function Names

Placing an integer superscript after the name or symbol of a function, as if the function were being raised to a power, commonly refers to repeated function composition rather than repeated multiplication. Thus $f^3(x)$ may mean $f(f(f(x)))$; in particular, $f^{-1}(x)$ usually denotes the inverse function of f. Iterated functions are of interest in the study of fractals and dynamical systems. Babbage was the first to study the problem of finding a functional square root $f^{1/2}(x)$.

However, for historical reasons, a special syntax applies to the trigonometric functions: a positive exponent applied to the function's abbreviation means that the result is raised to that power, while an exponent of –1 denotes the inverse function. That is, $\sin^2 x$ is just a shorthand way to write $(\sin x)^2$ without using parentheses, whereas $\sin^{-1}x$ refers to the inverse function of the sine, also called $\arcsin x$. There is no need for a shorthand for the reciprocals of trigonometric functions since each has its own name and abbreviation; for example, $1/(\sin x) = (\sin x)^{-1} = \csc x$. A similar convention applies to logarithms, where $\log^2 x$ usually means $(\log x)^2$, not $\log \log x$.

Generalizations

In Abstract Algebra

Exponentiation for integer exponents can be defined for quite general structures in abstract algebra.

Let X be a set with a power-associative binary operation which is written multiplicatively. Then x^n is defined for any element x of X

and any nonzero natural number n as the product of n copies of x, which is recursively defined by

$$\mathbb{R}\begin{cases} x^1 = x \\ x^n = x^{n-1}x \quad \text{for } n > 1 \end{cases}$$

One has the following properties

$$(x^i x^j)x^k = x^i(x^j x^k) \quad \text{(power-associative property)}$$
$$x^{m+n} = x^m x^n$$
$$(x^m)^n = x^{mn}$$

If the operation has a two-sided identity element 1 (often denoted by e), then x^0 is defined to be equal to 1 for any x.

$$x1 = 1x = x \quad \text{(two-sided identity)}$$
$$x^0 = 1$$

If the operation also has two-sided inverses, and multiplication is associative then the magma is a group. The inverse of x can be denoted by x^{-1} and follows all the usual rules for exponents.

$$xx^{-1} = x^{-1}x = 1 \quad \text{(two-sided inverse)}$$
$$(xy)z = x(yz) \quad \text{(associative)}$$
$$x^{-n} = \left(x^{-1}\right)^n$$
$$x^{m-n} = x^m x^{-n}$$

If the multiplication operation is commutative (as for instance in abelian groups), then the following holds:

$$(xy)^n = x^n y^n$$

If the binary operation is written additively, as it often is for abelian groups, then "exponentiation is repeated multiplication" can be reinterpreted as "multiplication is repeated addition". Thus, each of the laws of exponentiation above has an analogue among laws of multiplication.

When one has several operations around, any of which might be repeated using exponentiation, it is common to indicate which operation is being repeated by placing its symbol in the superscript. Thus, x^{-n} is $x* \cdots *x$, while $x^{\#n}$ is $x \# \cdots \# x$, whatever the operations " and # might be.

Superscript notation is also used, especially in group theory, to indicate conjugation. That is, $g^h = h^{-1}gh$, where g and h are elements of some group. Although conjugation obeys some of the same laws as exponentiation, it is not an example of repeated multiplication in any sense. A quandle is an algebraic structure in which these laws of conjugation play a central role.

Over Sets

If n is a natural number and A is an arbitrary set, the expression A^n is often used to denote the set of ordered n–tuples of elements of A. This is equivalent to letting A^n denote the set of functions from the set $\{0, 1, 2, \ldots, n-1\}$ to the set A; the n–tuple $(a_0, a_1, a_2, \ldots, a_{n-1})$ represents the function that sends i to a_i.

For an infinite cardinal number κ and a set A, the notation A^κ is also used to denote the set of all functions from a set of size κ to A. This is sometimes written ${}^\kappa A$ to distinguish it from cardinal exponentiation, defined below.

This generalized exponential can also be defined for operations on sets or for sets with extra structure. For example, in linear algebra, it makes sense to index direct sums of vector spaces over arbitrary index sets. That is, we can speak of

$$\bigoplus_{i \in \mathbb{N}} V_i$$

where each V_i is a vector space.

Then if $V_i = V$ for each i, the resulting direct sum can be written in exponential notation as $V^{\oplus N}$, or simply V^N with the understanding that the direct sum is the default. We can again replace the set N with a cardinal number n to get V^n, although without choosing a specific standard set with cardinality n, this is defined only up toisomorphism. Taking V to be the field R of real numbers (thought of as a vector space over itself) and n to be some natural number, we get the vector space that is most commonly studied in linear algebra, the Euclidean space R^n.

If the base of the exponentiation operation is a set, the exponentiation operation is the Cartesian product unless otherwise stated. Since multiple Cartesian products produce an n-tuple, which can be represented by a function on a set of appropriate cardinality, S^N becomes simply the set of all functions from N to S in this case:

$$S^N \equiv \{f : N \to S\}$$

This fits in with the exponentiation of cardinal numbers, in the sense that $|S^N| = |S|^{|N|}$, where $|X|$ is the cardinality of X. When "2" is defined as $\{0, 1\}$, we have $|2^X| = 2^{|X|}$, where 2^X, usually denoted by P(X), is the power set of X; each subset Y of X corresponds uniquely to a function on X taking the value 1 for $x \in Y$ and 0 for $x \notin Y$.

In Category Theory

In a Cartesian closed category, the exponential operation can be used to raise an arbitrary object to the power of another object.

This generalizes the Cartesian product in the category of sets. If 0 is an initial object in a Cartesian closed category, then the exponential object 0^0 is isomorphic to any terminal object 1.

Of Cardinal and Ordinal Numbers

In set theory, there are exponential operations for cardinal and ordinal numbers.

If κ and λ are cardinal numbers, the expression κ^λ represents the cardinality of the set of functions from any set of cardinality λ to any set of cardinality κ. If κ and λ are finite, then this agrees with the ordinary arithmetic exponential operation.

For example, the set of 3-tuples of elements from a 2-element set has cardinality $8 = 2^3$.

Exponentiation of cardinal numbers is distinct from exponentiation of ordinal numbers, which is defined by a limit process involving transfinite induction.

Repeated Exponentiation

Just as exponentiation of natural numbers is motivated by repeated multiplication, it is possible to define an operation based on repeated exponentiation; this operation is sometimes called tetration. Iterating tetration leads to another operation, and so on.

This sequence of operations is expressed by the Ackermann function and Knuth's up-arrow notation. Just as exponentiation grows faster than multiplication, which is faster growing than addition, tetration is faster growing than exponentiation.

Evaluated at (3,3), the functions addition, multiplication, exponentiation, tetration yield 6, 9, 27, and 7,625,597,484,987 respectively.

In Programming Languages

The superscript notation x^y is convenient in handwriting but inconvenient for typewriters and computer terminals that align the baselines of all characters on each line.

Many programming languages have alternate ways of expressing exponentiation that do not use superscripts:

- $x \uparrow y$: Algol, Commodore BASIC
- x ^ y: BASIC, J, MATLAB, R, Microsoft Excel, TeX (and its derivatives), TI-BASIC, bc (for integer exponents), Haskell (for nonnegative integer exponents), Lua, ASP and most computer algebra systems
- x ^^ y: Haskell (for fractional base, integer exponents), D
- x ** y: Ada, Bash, COBOL, Fortran, FoxPro, Gnuplot, OCaml, Perl, PL/I, Python, Rexx, Ruby, SAS, Seed7, Tcl, ABAP, Haskell (for floating-point exponents), Turing, VHDL
- $x_\star y$: APL

Many programming languages lack syntactic support for exponentiation, but provide library functions.

In Bash, C, C++, C#, Java, JavaScript, Perl, PHP, Python and Ruby, the symbol ^ represents bitwise XOR.

In Pascal, it represents indirection.

In OCaml and Standard ML, it represents string concatenation.

History of the Notation

The term *power* was used by the Greek mathematician Euclid for the square of a line. Archimedes discovered and proved the law of exponents, $10^a\, 10^b = 10^{a+b}$, necessary to manipulate powers of 10.

In the 9th century, the Persian mathematician Muhammad ibn Mûsâ al-Khwârizmî used the terms *mal* for a square and *kab* for a cube, which later Islamic mathematicians represented in mathematical notation as *m* and *k*, respectively, by the 15th century, as seen in the work of Abû al-Hasan ibn Alî al-Qalasâdî.

Nicolas Chuquet used a form of exponential notation in the 15th century, which was later used by Henricus Grammateus and Michael Stifel in the 16th century. Samuel Jeake introduced the term *indices* in 1696.

In the 16th century Robert Recorde used the terms square, cube, zenzizenzic (fourth power), sursolid (fifth), zenzicube (sixth), second sursolid (seventh), and zenzizenzizenzic (eighth). *Biquadrate* has been used to refer to the fourth power as well.

Some mathematicians (e.g., Isaac Newton) used exponents only for powers greater than two, preferring to represent squares as repeated multiplication. Thus they would write polynomials, for example, as $ax + bxx + cx^3 + d$.

Another historical synonym, involution, is now rare and should not be confused with its more common meaning.

Rational Function

In mathematics, a rational function is any function which can be written as the ratio of two polynomial functions.

Neither the coefficients of the polynomials nor the values taken by the function are necessarily rational numbers.

Definitions

In the case of one variable, x, a function is called a rational function if and only if it can be written in the form

$$f(x) = \frac{P(x)}{Q(x)}$$

where P and Q are polynomial functions in x and Q is not the zero polynomial.

The domain of f is the set of all points for which the denominator $Q(x)$ is not zero, where one assumes that the fraction is written in its lower degree terms, that is, P and Q have several factors of the positive degree.

Every polynomial function is a rational function with $Q(x) = 1$. A function that cannot be written in this form (for example, $f(x) = \sin(x)$) is not a rational function (but the adjective "irrational" is not generally used for functions, but only for numbers).

An expression of the form $\frac{P(x)}{Q(x)}$ is called a *rational expression*. The need not be a variable. In abstract algebra the x is called an *indeterminate*.

A *rational equation* is an equation in which two rational expressions are set equal to each other.

These expressions obey the same rules as fractions. The equations can be solved by cross-multiplying.

Division by zero is undefined, so that a solution causing formal division by zero is rejected.

Examples

Examples of rational functions:

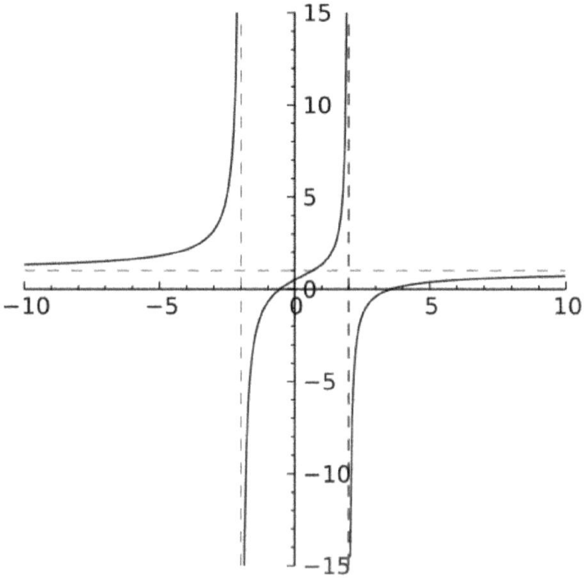

Figure: *Rational function of degree 2:* $y = \dfrac{x^2 - 3x - 2}{x^2 - 4}$

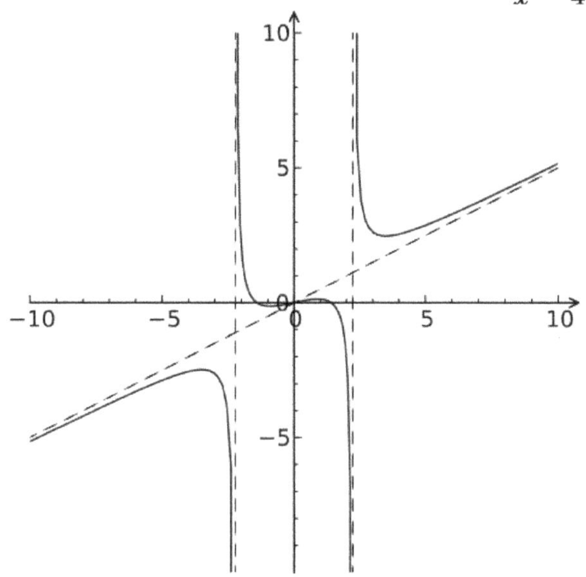

Figure: *Rational function of degree 3:* $y = \dfrac{x^3 - 2x}{2(x^2 - 5)}$

The rational function $f(x)=\dfrac{x^3-2x}{2(x^2-5)}$ is not defined at $x^2=5 \Leftrightarrow x=\pm\sqrt{5}$.

The rational function $f(x)=\dfrac{x^2+2}{x^2+1}$ is defined for all real numbers, but not for all complex numbers, since if x were a square root of – 1 (i.e. the imaginary unit or its negative), then formal evaluation would lead to division by zero: $f(i)=\dfrac{i^2+2}{i^2+1}=\dfrac{-1+2}{-1+1}=\dfrac{1}{0}$, which is undefined.

The rational function $f(x)=\dfrac{x^3-2x}{2(x^2-5)}$, as x approaches infinity, is asymptotic to $\dfrac{x}{2}$. A constant function such as $f(x) = \pi$ is a rational function since constants are polynomials. Note that the function itself is rational, even though $f(x)$ is irrational for all x.

The rational function $f(x)=\dfrac{x}{x}$ is equal to 1 for all x except 0, where there is a removable singularity.

The sum, product, or quotient (excepting division by the zero polynomial) of two rational functions is itself a rational function: however, the process of reduction to standard form may inadvertently result in the removing of such discontinuities unless care is taken.

Taylor Series

The coefficients of a Taylor series of any rational function satisfy a linear recurrence relation, which can be found by setting the rational function equal to its Taylor series and collecting like terms.

For example,

$$\frac{1}{x^2-x+2}=\sum_{k=0}^{\infty} a_k x^k.$$

Multiplying through by the denominator and distributing,

$$1=(x^2-x+2)\sum_{k=0}^{\infty} a_k x^k$$

$$1=\sum_{k=0}^{\infty} a_k x^{k+2}-\sum_{k=0}^{\infty} a_k x^{k+1}+2\sum_{k=0}^{\infty} a_k x^k.$$

After adjusting the indices of the sums to get the same powers of x, we get

$$1=\sum_{k=2}^{\infty}a_{k-2}x^k-\sum_{k=1}^{\infty}a_{k-1}x^k+2\sum_{k=0}^{\infty}a_kx^k.$$

Combining like terms gives

$$1=2a_0+(2a_1-a_0)x+\sum_{k=2}^{\infty}(a_{k-2}-a_{k-1}+2a_k)x^k.$$

Since this holds true for all x in the radius of convergence of the original Taylor series, we can compute as follows. Since the constant term on the left must equal the constant term on the right it follows that

$$a_0=\frac{1}{2}.$$

Then, since there are no powers of x on the left, all of the coefficients on the right must be zero, from which it follows that

$$a_1=\frac{1}{4}$$

$$a_k=\frac{1}{2}(a_{k-1}-a_{k-2})\quad \textit{for } k\geq 2.$$

Conversely, any sequence that satisfies a linear recurrence determines a rational function when used as the coefficients of a Taylor series. This is useful in solving such recurrences, since by using partial fraction decomposition we can write any rational function as a sum of factors of the form $1/(ax+b)$ and expand these as geometric series, giving an explicit formula for the Taylor coefficients; this is the method of generating functions.

Abstract Algebra and Geometric Notion

In abstract algebra the concept of a polynomial is extended to include formal expressions in which the coefficients of the polynomial can be taken from any field. In this setting given a field F and some indeterminate X, a rational expression is any element of the field of fractions of the polynomial ring $F[X]$.

Any rational expression can be written as the quotient of two polynomials P/Q with $Q \neq 0$, although this representation isn't unique. P/Q is equivalent to R/S, for polynomials P, Q, R, and S, when $PS = QR$. However since $F[X]$ is a unique factorization domain, there is a unique representation for any rational expression P/Q with P and

Q polynomials of lowest degree and Q chosen to be monic. This is similar to how a fraction of integers can always be written uniquely in lowest terms by canceling out common factors.

The field of rational expressions is denoted $F(X)$. This field is said to be generated (as a field) over F by (a transcendental element) X, because $F(X)$ does not contain any proper subfield containing both F and the element X.

Complex Rational Functions

In complex analysis, a rational function

$$f(z)=\frac{P(z)}{Q(z)}$$

is the ratio of two polynomials with complex coefficients, where Q is not the zero polynomial and P and Q have no common factor (this avoids f taking the indeterminate value 0/0). The domain and range of f are usually taken to be the Riemann sphere, which avoids any need for special treatment at the poles of the function (where $Q(z)$ is 0).

The *degree* of a rational function is the maximum of the degrees of its constituent polynomials P and Q. If the degree of f is d, then the equation

$$f(z)=w$$

has d distinct solutions in z except for certain values of w, called *critical values*, where two or more solutions coincide. The function f can therefore be thought of as a d-fold covering of the w-sphere by the z-sphere.

Rational functions with degree 1 are called *Möbius transformations* and form the automorphisms group of the Riemann sphere. Rational functions are representative examples of meromorphic functions.

Notion of a Rational Function on an Algebraic Variety

Like polynomials, rational expressions can also be generalized to n indeterminates $X_1,\ldots, X_n$, by taking the field of fractions of $F[X_1,\ldots, X_n]$, which is denoted by $F(X_1,\ldots, X_n)$.

An extended version of the abstract idea of rational function is used in algebraic geometry. There the function field of an algebraic variety V is formed as the field of fractions of the coordinate ring of V (more accurately said, of a Zariski-dense affine open set in V). Its elements f are considered as regular functions in the sense of algebraic geometry on non-empty open sets U, and also may be seen as morphisms to the projective line.

Applications

These objects are first encountered in school algebra. In more advanced mathematics they play an important role in ring theory, especially in the construction of field extensions. They also provide an example of a *nonarchimedean field.*

Rational functions are used in numerical analysis for interpolation and approximation of functions, for example the Padé approximations introduced by Henri Padé. Approximations in terms of rational functions are well suited for computer algebra systems and other numerical software. Like polynomials, they can be evaluated straightforwardly, and at the same time they express more diverse behaviour than polynomials.

Rational functions are used to approximate or model more complex equations in science and engineering including:

(*i*) fields and forces in physics,

(*ii*) spectroscopy in analytical chemistry,

(*iii*) enzyme kinetics in biochemistry,

(*iv*) electronic circuitry,

(*v*) aerodynamics,

(*vi*) medicine concentrations in vivo,

(*vii*) wave functions for atoms and molecules,

(*viii*) optics and photography to improve image resolution, and

(*ix*) acoustics and sound.

Polynomial

In mathematics, a polynomial is an expression of finite length constructed from variables (also called indeterminates) and constants, using only the operations of addition, subtraction, multiplication, and non-negative integer exponents. However, the division by a constant is allowed, because the multiplicative inverse of a non zero constant is also a constant. For example, $x^2 - x/4 + 7$ is a polynomial, but $x^2 - 4/x + 7x^{3/2}$ is an algebraic expression that is not a polynomial, because its second term involves a division by the variable x (the term $4/x$), and also because its third term contains an exponent that is not a non-negative integer (3/2). The term "polynomial" can also be used as an adjective, for quantities that can be expressed as a polynomial of some parameter, as in *polynomial time,* which is used in computational complexity theory.

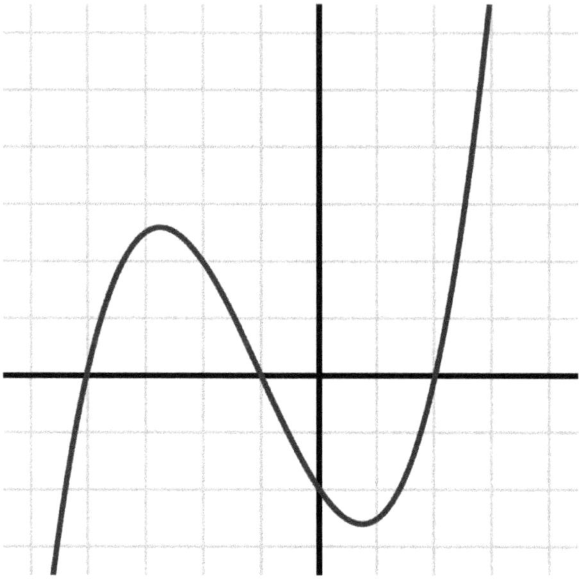

Figure: *The graph of a polynomial function of degree 3.*

Polynomial comes from the Greek *poly*, "many" and medieval Latin *binomium*, "binomial". The word was introduced in Latin by Franciscus Vieta.

Polynomials appear in a wide variety of areas of mathematics and science. For example, they are used to form polynomial equations, which encode a wide range of problems, from elementary word problems to complicated problems in the sciences; they are used to define polynomial functions, which appear in settings ranging from basic chemistry and physics to economics and social science; they are used in calculus and numerical analysis to approximate other functions. In advanced mathematics, polynomials are used to construct polynomial rings, a central concept in abstract algebra and algebraic geometry.

Overview

A polynomial is either zero or can be written as the sum of a finite number of non-zero terms. Each term consists of the product of a constant (called the coefficient of the term) and a finite number of variables (usually represented by letters), also called indeterminates, raised to whole number powers. The exponent on a variable in a term is called the degree of that variable in that term; the degree of the term is the sum of the degrees of the variables in that term, and the degree of a polynomial is the largest degree of any one term. Since $x = x^1$, the degree of a variable without a written exponent is one. A term with no variables is called a constant term, or just a constant; the degree of a (nonzero) constant term is 0. The coefficient of a term may be any number from a specified set. If that set is the set of real numbers, we speak of "polynomials over the reals". Other common kinds of polynomials are polynomials with integer coefficients, polynomials with complex coefficients, and polynomials with coefficients that are integers modulo of some prime number p. In most of the examples in this section, the coefficients are integers.

For Example:

$$-5x^2y$$

is a term. The coefficient is –5, the variables are x and y, the degree of x is in the term two, while the degree of y is one.

The degree of the entire term is the sum of the degrees of each variable in it, so in this example the degree is 2 + 1 = 3.

Forming a sum of several terms produces a polynomial. For example, the following is a polynomial:

$$\underbrace{3x^2}_{\substack{\text{term}\\1}}\underbrace{-5x}_{\substack{\text{term}\\2}}\underbrace{+4}_{\substack{\text{term}\\3}}.$$

It consists of three terms: the first is degree two, the second is degree one, and the third is degree zero.

The commutative law of addition can be used to rearrange terms into any preferred order. In polynomials with one variable, the terms are usually ordered according to degree, either in "descending powers of x", with the term of largest degree first, or in "ascending powers of x". The polynomial in the example above is written in descending powers of x. The first term has coefficient 3, variable x, and exponent 2. In the second term, the coefficient is –5. The third term is a constant. Since the degree of a non-zero polynomial is the largest degree of any one term, this polynomial has degree two.

Two terms with the same variables raised to the same powers are called "similar terms" or "like terms", and they can be combined, using the distributive law, into a single term whose coefficient is the sum of the coefficients of the terms that were combined. It may happen that this makes the coefficient 0. Polynomials can be added using the associative law of addition (grouping all their terms together into a single sum), possibly followed by reordering, and combining of like terms. For example, if

$$P = 3x^2 - 2x + 5xy - 2$$

$$Q = -3x^2 + 3x + 4y^2 + 8,$$

then

$$P + Q = 3x^2 - 2x + 5xy - 2 - 3x^2 + 3x + 4y^2 + 8,$$

which can be simplified to

$$P + Q = x + 5xy + 4y^2 + 6.$$

To work out the product of two polynomials into a sum of terms, the distributive law is repeatedly applied, which results in each term of one polynomial being multiplied by every term of the other. For example, if

$$P = 2x + 3y + 5$$

$$Q = 2x + 5y + xy + 1,$$

then

$$\begin{array}{rccccccc} PQ = & (2x\cdot 2x) & + & (2x\cdot 5y) & + & (2x\cdot xy) & + & (2x\cdot 1) \\ + & (3y\cdot 2x) & + & (3y\cdot 5y) & + & (3y\cdot xy) & + & (3y\cdot 1) \\ + & (5\cdot 2x) & + & (5\cdot 5y) & + & (5\cdot xy) & + & (5\cdot 1) \end{array}$$

which can be simplified to

$$PQ = 4x^2 + 21xy + 2x^2y + 12x + 15y^2 + 3xy^2 + 28y + 5.$$

The sum or product of two polynomials is always a polynomial.

Alternative Forms

In general any expression can be considered a polynomial if it is built from variables and constants using only addition, subtraction, multiplication, and raising expressions to whole number powers. Such an expression can always be rewritten as a sum of terms. For example, $(x + 1)^3$ is a polynomial; its standard form is $x^3 + 3x^2 + 3x + 1$. However, in some situations, a more accurate terminology is needed in order to not be confusing.

In such a case, one should say: $(x + 1)^3$ is a *polynomial expression* which may be *expanded* or *rewritten* into the *polynomial* $x^3 + 3x^2 + 3x + 1$. Although different as expressions, these two expressions are *equal* in the ring of the polynomials in the indeterminate x with integer coefficients.

As for the integers, two kinds of divisions are considered for the polynomials. The *Euclidean division* that generalizes the Euclidean division of the integers. It results in two polynomials, a *quotient* and a *remainder* that are characterized by the following property of the polynomials: given two polynomials a and b such that $b \neq 0$, there exists a unique pair of polynomials, q, the quotient, and r, the remainder, such that $a = bq + r$ and degree(r) < degree(b) (here the polynomial zero is supposed to have a negative degree). By hand as well as with a computer, this division can be computed by the polynomial long division algorithm.

A formal quotient of polynomials, that is, an algebraic fraction where the numerator and denominator are polynomials, is called a "rational expression" or "rational fraction" and is not, in general, a polynomial. Division of a polynomial by a number, however, does yield another polynomial. For example, $x^3/12$ is considered a valid term in a polynomial (and a polynomial by itself) because it is equivalent to $(1/12)x^3$ and 1/12 is just a constant. When this expression is used as a term, its coefficient is therefore 1/12. For similar reasons, if complex coefficients are allowed, one may have a single term like $(2 + 3i)\,x^3$; even though it looks like it should be expanded to two terms, the complex number $2 + 3i$ is one complex number, and is the coefficient of that term. The expression $1/(x^2 + 1)$ is not a polynomial because it includes division by a non-constant polynomial. The expression $(5 + y)^x$ is not a polynomial, because it contains a variable used as exponent.

Since subtraction can be replaced by addition of the opposite quantity, and since positive whole number exponents can be replaced by repeated multiplication, all polynomials can be constructed from constants and variables using only addition and multiplication.

Polynomial Functions

A polynomial function is a function that can be defined by evaluating a polynomial. A function f of one argument is called a polynomial function if it satisfies

$$f(x) = a_n x^n + a_{n-1} x^{n-1} + \cdots + a_2 x^2 + a_1 x + a_0$$

for all arguments x, where n is a non-negative integer and a_0, a_1, a_2, ..., a_n are constant coefficients.

For example, the function f, taking real numbers to real numbers, defined by

$$f(x) = x^3 - x$$

is a polynomial function of one argument. Polynomial functions of multiple arguments can also be defined, using polynomials in multiple variables, as in

$$f(x,y) = 2x^3 + 4x^2y + xy^5 + y^2 - 7.$$

An example is also the function $f(x) = \cos(2\arccos(x))$ which, although it doesn't look like a polynomial, is a polynomial function since for every x it is true that $f(x) = 2x^2 - 1$.

Polynomial functions are a class of functions having many important properties. They are all continuous, smooth, entire, computable, etc.

Polynomial Equations

A polynomial equation, also called algebraic equation, is an equation in which a polynomial is set equal to another polynomial.

$$3x^2 + 4x - 5 = 0$$

is a polynomial equation. In case of a univariate polynomial equation, the variable is considered an unknown, and one seeks to find the possible values for which both members of the equation evaluate to the same value (in general more than one solution may exist). A polynomial equation stands in contrast to a polynomial identity like $(x + y)(x - y) = x^2 - y^2$, where both members represent the same polynomial in different forms, and as a consequence any evaluation of both members gives a valid equality. This means that a polynomial identity is a polynomial equation for which all possible values of the unknowns are solutions.

Elementary Properties of Polynomials

- A sum of polynomials is a polynomial.
- A product of polynomials is a polynomial.
- A composition of two polynomials is a polynomial, which is obtained by substituting a variable of the first polynomial by the second polynomial.
- The derivative of the polynomial $a_nx^n + a_{n-1}x^{n-1} + ... + a_2x^2 + a_1x + a_0$ is the polynomial $na_nx^{n-1} + (n\text{-}1)a_{n-1}x^{n-2} + ... + 2a_2x + a_1$. If the set of the coefficients does not contain the integers (for example if the coefficients are integers modulo some prime number p), then ka_k should be interpreted as the sum of a_k with itself, k times. For example, over the integers modulo p, the derivative of the polynomial x^p+1 is the polynomial 0.

- If the division by integers is allowed in the set of coefficients, a primitive or anti-derivative of the polynomial $a_n x^n + a_{n-1}x^{n-1} + \ldots + a_2x^2 + a_1x + a_0$ is $a_n x^{n+1}/(n+1) + a_{n-1}x^n/n + \ldots + a_2x^3/3 + a_1x^2/2 + a_0x + c$, where c is an arbitrary constant. Thus x^2+1 is a polynomial with integer coefficients whose primitives are not polynomials over the integers. If this polynomial is viewed as a polynomial over the integers modulo 3 it has no primitive at all.

Polynomials serve to approximate other functions, such as sine, cosine, and exponential.

All polynomials have an expanded form, in which the distributive and associative laws have been used to remove all brackets and commutative law has been used to make the like terms adjacent and combine them. All polynomials with coefficients in a unique factorization domain (for example, the integers or a field) also have a factored form in which the polynomial is written as a product of irreducible polynomials and a constant. In the case of the field of complex numbers, the irreducible polynomials are linear. For example, the factored form of

$$5x^3 - 5$$

is

$$5(x-1)(x^2+x+1),$$

over the integers and

$$5(x-1)\left(x+\frac{1+i\sqrt{3}}{2}\right)\left(x+\frac{1-i\sqrt{3}}{2}\right)$$

over the complex numbers.

Every polynomial in one variable is equivalent to a polynomial with the form

$$a_n x^n + a_{n-1}x^{n-1} + \cdots + a_2x^2 + a_1x + a_0.$$

This form is sometimes taken as the definition of a polynomial in one variable.

Evaluation of a polynomial consists of assigning a number to each variable and carrying out the indicated multiplications and additions. Actual evaluation is usually more efficient using the Horner scheme:

$$((\cdots((a_n x + a_{n-1})x + a_{n-2})x + \cdots + a_3)x + a_2)x + a_1)x + a_0.$$

Polynomial evaluation can also be accomplished through polynomial division, since the remainder of $f(x)/(x–a)$ is $f(a)$.

In elementary algebra, methods are given for solving all first degree and second degree polynomial equations in one variable. In the case of polynomial equations, the variable is often called an *unknown*. The number of solutions may not exceed the degree, and equals the degree when multiplicity of solutions and complex number solutions are counted. This fact is called the fundamental theorem of algebra.

A system of polynomial equations is a set of equations in which each variable must take on the same value everywhere it appears in any of the equations. Systems of equations are usually grouped with a single open brace on the left. In elementary algebra, in particular in linear algebra, methods are given for solving a system of linear equations in several unknowns. If there are more unknowns than equations, the system is called underdetermined. If there are more equations than unknowns, the system is called overdetermined. Overdetermined systems are common in practical applications. For example, one U.S. mapping survey used computers to solve 2.5 million equations in 400,000 unknowns.

Viète's formulas relate the coefficients of a polynomial to symmetric polynomial functions of its roots.

History

Determining the roots of polynomials, or "solving algebraic equations", is among the oldest problems in mathematics. However, the elegant and practical notation we use today only developed beginning in the 15th century. Before that, equations were written out in words. For example, an algebra problem from the Chinese Arithmetic in Nine Sections, circa 200 BCE, begins "Three sheafs of good crop, two sheafs of mediocre crop, and one sheaf of bad crop are sold for 29 dou." We would write $3x + 2y + z = 29$.

Notation

The earliest known use of the equal sign is in Robert Recorde's *The Whetstone of Witte*, 1557. The signs + for addition, – for subtraction, and the use of a letter for an unknown appear in Michael Stifel's *Arithemetica integra*, 1544. René Descartes, in *La géometrie*, 1637, introduced the concept of the graph of a polynomial equation. He popularized the use of letters from the beginning of the alphabet to denote constants and letters from the end of the alphabet to denote variables, as can be seen above, in the general formula for a polynomial in one variable, where the a 's denote constants and x denotes a variable. Descartes introduced the use of superscripts to denote exponents as well.

Solving Polynomial Equations

Every polynomial P in x corresponds to a function, $f(x) = P$ (where the occurrences of x in P are interpreted as the argument of f), called the *polynomial function* of P; the equation in x setting $f(x) = 0$ is the *polynomial equation* corresponding to P. The solutions of this equation are called the *roots* of the polynomial; they are the *zeroes* of the function f (corresponding to the points where the graph of f meets the x-axis). A number a is a root of P if and only if the polynomial $x - a$ (of degree one in x) divides P. It may happen that $x - a$ divides P more than once: if $(x - a)^2$ divides P then a is called a multiple root of P, and otherwise a is called a simple root of P. If P is a nonzero polynomial, there is a highest power m such that $(x - a)^m$ divides P, which is called the *multiplicity* of the root a in P. When P is the zero polynomial, the corresponding polynomial equation is trivial, and this case is usually excluded when considering roots: with the above definitions every number would be a root of the zero polynomial, with undefined (or infinite) multiplicity. With this exception made, the number of roots of P, even counted with their respective multiplicities, cannot exceed the degree of P.

Some polynomials, such as $x^2 + 1$, do not have any roots among the real numbers. If, however, the set of allowed candidates is expanded to the complex numbers, every non-constant polynomial has at least one root; this is the fundamental theorem of algebra. By successively dividing out factors $x - a$, one sees that any polynomial with complex coefficients can be written as a constant (its leading coefficient) times a product of such polynomial factors of degree 1; as a consequence the number of (complex) roots counted with their multiplicities is exactly equal to the degree of the polynomial.

There is a difference between approximating roots and finding exact expressions for roots. Formulas for expressing the roots of polynomials of degree 2 in terms of square roots have been known since ancient times, and for polynomials of degree 3 or 4 similar formulas (using cube roots in addition to square roots) were found in the 16th century. But formulas for degree 5 eluded researchers. In 1824, Niels Henrik Abel proved the striking result that there can be no general (finite) formula, involving only arithmetic operations and radicals, that expresses the roots of a polynomial of degree 5 or greater in terms of its coefficients. In 1830, Évariste Galois, studying the permutations of the roots of a polynomial, extended Abel-Ruffini theorem by showing that, given a polynomial equation, one may decide if it is solvable by radicals, and, if it is, solve it. This result marked the start of Galois theory and Group theory, two important branches of modern mathematics. Galois himself noted that the computations implied by his method were impracticable. Nevertheless formulas for solvable equations of degrees 5 and 6 have been published.

Numerical approximations of roots of polynomial equations in one unknown is easily done on a computer by the Jenkins-Traub method, Laguerre's method, Durand–Kerner method or by some other root-finding algorithm. For polynomials in more than one variable the notion of root does not exist, and there are usually infinitely many combinations of values for the variables for which the polynomial function takes the value zero. However for certain *sets* of such polynomials it may happen that for only finitely many combinations all polynomial functions take the value zero.

For a set of polynomial equations in several unknowns, there are algorithms to decide if they have a finite number of complex solutions. If the number of solutions is finite, there are algorithms to compute the solutions. The methods underlying these algorithms are described in the article systems of polynomial equations. The special case where all the polynomials are of degree one is called a system of linear equations, for which another range of different solution methods exist, including the classical Gaussian elimination.

It has been shown by Richard Birkeland and Karl Meyr that the roots of any polynomial may be expressed in terms of multivariate hypergeometric functions. Ferdinand von Lindemann and Hiroshi Umemura showed that the roots may also be expressed in terms of Siegel modular functions, generalizations of the theta functions that appear in the theory of elliptic functions. These characterisations of the roots of arbitrary polynomials are generalisations of the methods previously discovered to solve the quintic equation.

Graphs

A polynomial function in one real variable can be represented by a graph.

- The graph of the zero polynomial

  $f(x) = 0$

  is the x-axis.

- The graph of a degree 0 polynomial
  $f(x) = a_0$, where $a_0 \neq 0$,
  is a horizontal line with y-intercept a_0
- The graph of a degree 1 polynomial (or linear function)
  $f(x) = a_0 + a_1 x$, where $a_1 \neq 0$,
  is an oblique line with y-intercept a_0 and slopea_1.
- The graph of a degree 2 polynomial
  $f(x) = a_0 + a_1 x + a_2 x^2$, where $a_2 \neq 0$
  is a parabola.
- The graph of a degree 3 polynomial
  $f(x) = a_0 + a_1 x + a_2 x^2, + a_3 x^3$, where $a_3 \neq 0$
  is a cubic curve.
- The graph of any polynomial with degree 2 or greater
  $f(x) = a_0 + a_1 x + a_2 x^2 + \ldots + a_n x^n$,
  where $a_n \neq 0$ and $n \geq 2$ is a continuous non-linear curve.

The graph of a non-constant (univariate) polynomial always tends to infinity when the variable increases indefinitely (in absolute value). Polynomial graphs are analyzed in calculus using intercepts, slopes, concavity, and end behaviour. The illustrations below show graphs of polynomials.

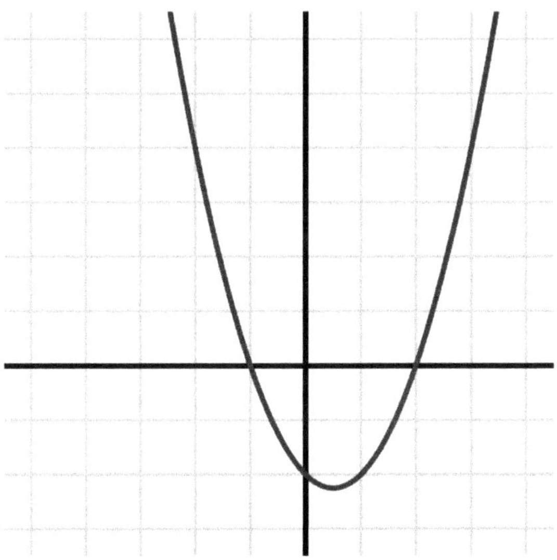

Figure: *Polynomial of degree 2:*

$f(x) = x^2 - x - 2 = (x+1)(x-2)$

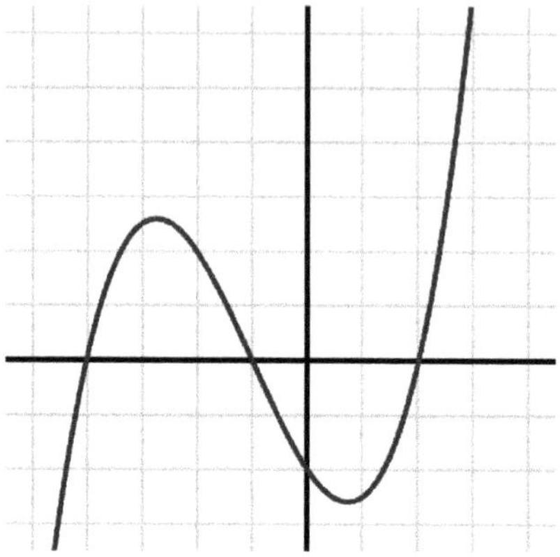

Figure: *Polynomial of degree 3:*

$f(x) = x^3/4 + 3x^2/4 - 3x/2 - 2 = 1/4\ (x+4)(x+1)(x-2)$

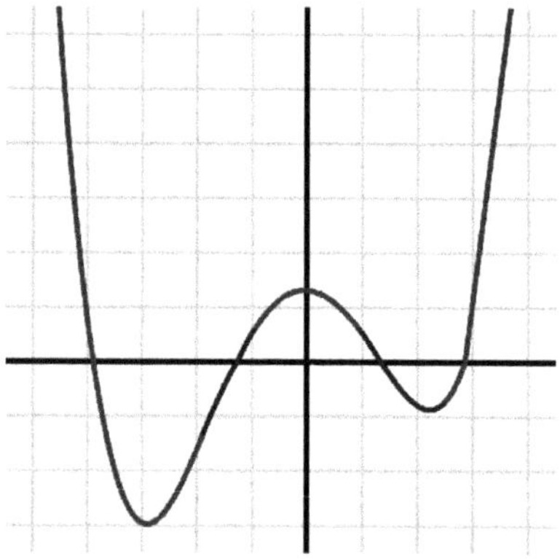

Figure: *Polynomial of degree 4:*

$f(x) = 1/14\ (x+4)(x+1)(x-1)(x-3) + 0.5$

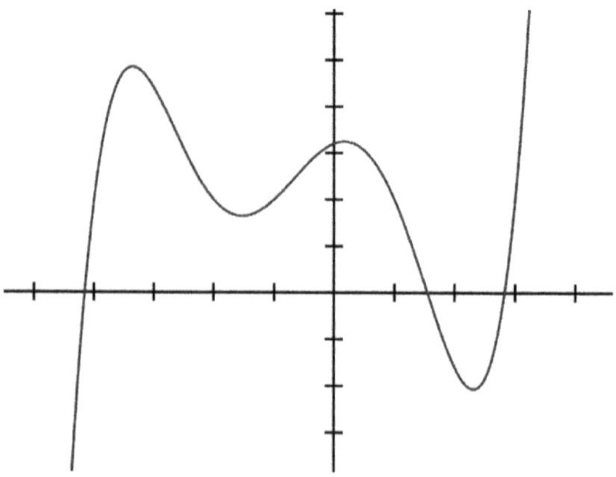

Figure: *Polynomial of degree 5:*

$f(x) = 1/20\ (x+4)(x+2)(x+1)(x-1)(x-3) + 2$

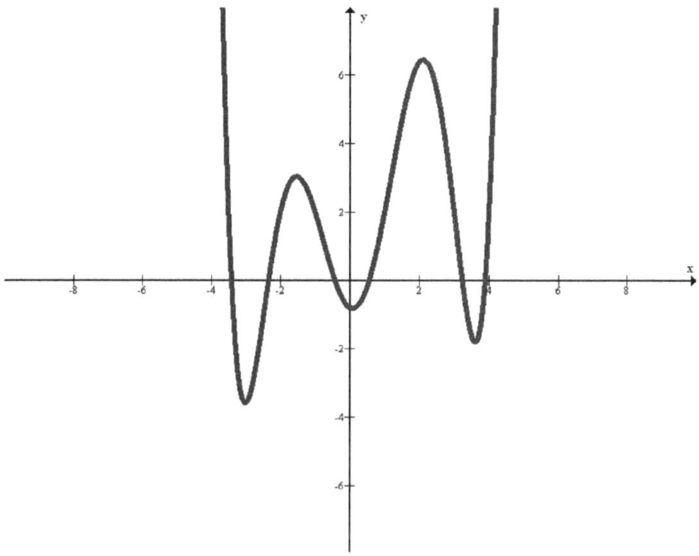

***Figure:** Polynomial of degree 6:*

$f(x) = 1/30\ (x+3.5)(x+2)(x+1)(x-1)(x-3)(x-4) + 2$

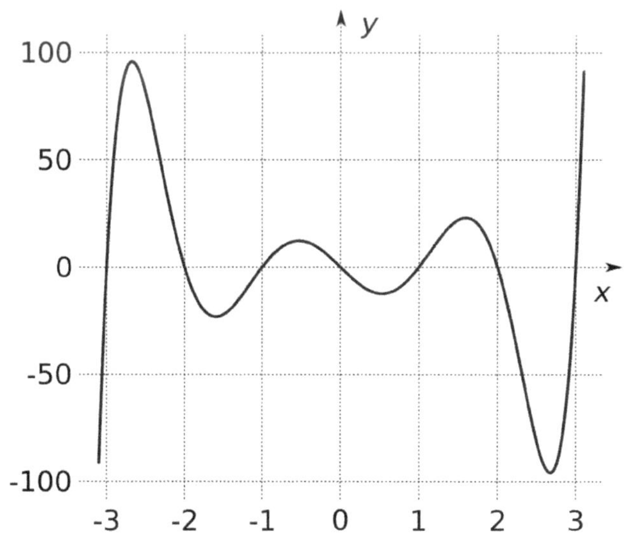

***Figure:** Polynomial of degree 7:*

$f(x) = (x-3)(x-2)(x-1)(x)(x+1)(x+2)(x+3)$

Polynomials and Calculus

One important aspect of calculus is the project of analyzing complicated functions by means of approximating them with polynomial functions. The culmination of these efforts is Taylor's theorem, which roughly states that every differentiable function locally looks like a polynomial function, and the Stone-Weierstrass theorem, which states that every continuous function defined on a compact interval of the real axis can be approximated on the whole interval as closely as desired by a polynomial function. Polynomial functions are also frequently used to interpolate functions.

Calculating derivatives and integrals of polynomial functions is particularly simple. For the polynomial function

$$\sum_{i=0}^{n} a_i x^i$$

the derivative with respect to x is

$$\sum_{i=1}^{n} a_i i x^{i-1}$$

and the indefinite integral is

$$\sum_{i=0}^{n} \frac{a_i}{i+1} x^{i+1} + c.$$

Abstract Algebra

In abstract algebra, one distinguishes between *polynomials* and *polynomial functions*. A polynomial f in one variable X over a ring R is defined as a formal expression of the form

$$f = a_n X^n + a_{n-1} X^{n-1} + \cdots + a_1 X^1 + a_0 X^0$$

where n is a natural number, the coefficients $a_0, \ldots, a_n$ are elements of R, and X is a formal symbol, whose powers X^i are just placeholders for the corresponding coefficients a_i, so that the given formal expression is just a way to encode the sequence $(a_0, a_1, \ldots)$, where there is an n such that $a_i = 0$ for all $i > n$. Two polynomials sharing the same value of n are considered equal if and only if the sequences of their coefficients are equal; furthermore any polynomial is equal to any polynomial with greater value of n obtained from it by adding terms in front whose coefficient is zero.

These polynomials can be added by simply adding corresponding coefficients (the rule for extending by terms with zero coefficients can be used to make sure such coefficients exist). Thus each polynomial is actually equal to the sum of the terms used in its formal expression, if such a term $a_i X^i$ is interpreted as a polynomial that has zero coefficients at all powers of X other than X^i. Then to define multiplication, it suffices by the distributive law to describe the product of any two such terms, which is given by the rule

$$aX^k\, bX^l = abX^{k+l}$$

for all elements a, b of the ring R and all natural numbers k and l.

Thus the set of all polynomials with coefficients in the ring R forms itself a ring, the *ring of polynomials* over R, which is denoted by $R[X]$. The map from R to $R[X]$ sending r to rX^0 is an injective homomorphism of rings, by which R is viewed as a subring of $R[X]$. If R is commutative, then $R[X]$ is an algebra over R.

One can think of the ring $R[X]$ as arising from R by adding one new element X to R, and extending in a minimal way to a ring in which X satisfies no other relations than the obligatory ones, plus commutation with all elements of R (that is $Xr = rX$). To do this, one must add all powers of X and their linear combinations as well.

Formation of the polynomial ring, together with forming factor rings by factoring out ideals, are important tools for constructing new rings out of known ones. For instance, the ring (in fact field) of complex numbers, which can be constructed from the polynomial ring $R[X]$ over the real numbers by factoring out the ideal of multiples of the polynomial $X^2 + 1$. Another example is the construction of finite fields, which proceeds similarly, starting out with the field of integers modulo some prime number as the coefficient ring R.

If R is commutative, then one can associate to every polynomial P in $R[X]$, a polynomial function f with domain and range equal to R (more generally one can take domain and range to be the same unital associative algebra over R). One obtains the value $f(r)$ by substitution of the value r for the symbol X in P. One reason to distinguish between polynomials and polynomial functions is that over some rings different polynomials may give rise to the same polynomial function.

This is not the case when R is the real or complex numbers, whence the two concepts are not always distinguished in analysis. An even more important reason to distinguish between polynomials and polynomial functions is that many operations on polynomials (like Euclidean division) require looking at what a polynomial is composed of as an expression rather than evaluating it at some constant value for X. And it should be noted that if R is not commutative, there is no (well behaved) notion of polynomial function at all.

Divisibility

In commutative algebra, one major focus of study is divisibility among polynomials. If R is an integral domain and f and g are polynomials in $R[X]$, it is said that f *divides* g or f is a divisor of g if there exists a polynomial q in $R[X]$ such that $fq = g$. One can show that every zero gives rise to a linear divisor, or more formally, if f is a polynomial in $R[X]$ and r is an element of R such that $f(r) = 0$, then the polynomial $(X - r)$ divides f. The converse is also true. The quotient can be computed using the polynomial long division.

If F is a field and f and g are polynomials in $F[X]$ with $g \neq 0$, then there exist unique polynomials q and r in $F[X]$ with

$$f = qg + r$$

and such that the degree of r is smaller than the degree of g (using the convention that the polynomial 0 has a negative degree). The polynomials q and r are uniquely determined by f and g. This is called Euclidean division, division with remainder or polynomial long division and shows that the ring $F[X]$ is a Euclidean domain.

Analogously, prime polynomials (more correctly, irreducible polynomials) can be defined as *polynomials which cannot be factorized into the product of two non constant polynomials*. Any polynomial may be decomposed into the product of a constant by a product of irreducible polynomials. This decomposition is unique up to the order of the factors and the multiplication of any constant factors by a constant (and division of the constant factor by the same constant. When the coefficients belong to a finite field or are rational numbers, there are algorithms to test irreducibility and to compute the factorization into irreducible polynomials. These algorithms are not practicable for hand written computation, but are available in any Computer algebra system (see Berlekamp's algorithm for the case in which the coefficients belong to a finite field or the Berlekamp–Zassenhaus algorithm when working over the rational numbers). Eisenstein's criterion can also be used in some cases to determine irreducibility.

Classifications

Polynomials are classified according to many different properties.

Number of Variables

One classification of polynomials is based on the number of distinct variables. A polynomial in one variable is called a univariate polynomial, a polynomial in more than one variable is called a multivariate polynomial. These notions refer more to the kind of polynomials one is generally working with than to individual polynomials; for instance when working with univariate polynomials one does not exclude constant polynomials (which may result, for instance, from the subtraction of non-constant polynomials), although strictly speaking constant polynomials do not contain any variables at all. It is possible to further classify multivariate polynomials as bivariate, trivariate, and so on, according to the maximum number of variables allowed. Again, so that the set of objects under consideration be closed under subtraction, a study of trivariate polynomials usually allows bivariate polynomials, and so on. It is common, also, to say simply "polynomials in x, y, and z", listing the variables allowed. In this case, xy is allowed.

Coefficients

Another classification of polynomials is by the kind of constant values allowed as coefficients. One can work with polynomials with integer, rational, real, or complex coefficients, and in abstract algebra polynomials with many other types of coefficients can be defined, such as integers modulo p. Especially interesting is the case of polynomials whose coefficients are polynomial in other variables. This allows us to consider a polynomial in, for example, x, y, z as a univariate polynomial in z with coefficients which are bivariate polynomials in x and y. This allows proofs by recursion on the number of variables and is frequently used in the theory of multivariate polynomials. As in the classification by number of variables, when working with coefficients for a given set, such as the complex numbers, coefficients from any subset are allowed. Thus $x^2 + 3x - 5$ is a polynomial with integer coefficients, but it is also a polynomial with complex coefficients, because the integers are a subset of the complex numbers.

Number of Non-zero Terms

Polynomials may also be classified by the number of terms with nonzero coefficients, so that a one-term polynomial is called a monomial, a two-term polynomial is called a binomial, and so on. (Some authors use "monomial" to mean "monic monomial".)

Polynomials Associated to Other Objects

Polynomials are frequently used to encode information about some other object. The characteristic polynomial of a matrix or linear operator contains information about the operator's eigenvalues. The minimal polynomial of an algebraic element records the simplest algebraic relation satisfied by that element. The chromatic polynomial of a graph counts the number of proper colourings of that graph.

Extensions of the Concept of a Polynomial

Polynomials can involve more than one variable, in which they are called multivariate. Rings of polynomials in a finite number of variables are of fundamental importance in algebraic geometry which studies the simultaneous zero sets of several such multivariate polynomials. These rings can alternatively be constructed by repeating the construction of univariate polynomials with as coefficient ring another ring of polynomials: thus the ring $R[X,Y]$ of polynomials in X and Y can be viewed as the ring $(R[X])[Y]$ of polynomials in Y with as coefficients polynomials in X, or as the ring $(R[Y])[X]$ of polynomials in X with as coefficients polynomials in Y. These identifications are compatible with arithmetic operations (they are isomorphisms of rings), but some notions such as degree or whether a polynomial is considered monic can change between these points of view. One can construct rings of polynomials in infinitely many variables, but since polynomials are (finite) expressions, any individual polynomial can only contain finitely many variables.

A binary polynomial where the second variable takes the form of an exponential function applied to the first variable, for example $P(X,e^X)$, may be called an exponential polynomial.

Laurent polynomials are like polynomials, but allow negative powers of the variable(s) to occur.

Quotients of polynomials are called rational expressions (or rational fractions), and functions that evaluate rational expressions are called rational functions. Rational fractions are formal quotients of polynomials (they are formed from polynomials just as rational numbers are formed from integers, writing a fraction of two of them; fractions related by the canceling of common factors are identified with each other). The rational function defined by a rational fraction is the quotient of the polynomial functions defined by the numerator and the denominator of the rational fraction. The rational fractions contain the Laurent polynomials, but do not limit denominators to powers of a variable. While polynomial functions are defined for all values of the variables, a rational function is defined only for the values of the variables for which the denominator is not null. A rational function produces rational output for any rational input for which it is defined; this is not true of other functions such as trigonometric functions, logarithms and exponential functions.

Formal power series are like polynomials, but allow infinitely many non-zero terms to occur, so that they do not have finite degree.

Unlike polynomials they cannot in general be explicitly and fully written down (just like real numbers cannot), but the rules for manipulating their terms are the same as for polynomials.

Cartesian Coordinate System

A Cartesian coordinate system is a coordinate system that specifies each point uniquely in a plane by a pair of numerical coordinates, which are the signed distances from the point to two fixed perpendicular directed lines, measured in the same unit of length. Each reference line is called a *coordinate axis* or just *axis* of the system, and the point where they meet is its *origin*, usually at ordered pair (0,0). The coordinates can also be defined as the positions of the perpendicular projections of the point onto the two axes, expressed as signed distances from the origin.

One can use the same principle to specify the position of any point in three-dimensional space by three Cartesian coordinates, its signed distances to three mutually perpendicular planes (or, equivalently, by its perpendicular projection onto three mutually perpendicular lines). In general, one can specify a point in a space of any dimension *n* by use of *n* Cartesian coordinates, the signed distances from *n* mutually perpendicular hyper planes.

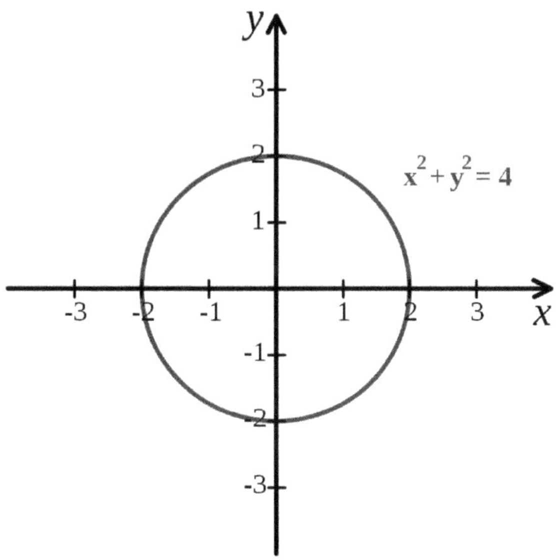

Figure: *Cartesian coordinate system with a circle of radius 2 entered at the origin marked in red. The equation of a circle is* $(x - a)^2 + (y - b)^2 = r^2$ *where a and b are the coordinates of the centre (a, b) and r is the radius.*

The invention of Cartesian coordinates in the 17th century by René Descartes (Latinized name: *Cartesius*) revolutionized mathematics by providing the first systematic link between Euclidean geometry and algebra. Using the Cartesian coordinate system, geometric shapes (such as curves) can be described by Cartesian equations: algebraic equations involving the coordinates of the points lying on the shape. For example, a circle of radius 2 may be described as the set of all points whose coordinates x and y satisfy the equation $x^2 + y^2 = 4$.

Cartesian coordinates are the foundation of analytic geometry, and provide enlightening geometric interpretations for many other branches of mathematics, such as linear algebra, complex analysis, differential geometry, multivariate calculus, group theory, and more. A familiar example is the concept of the graph of a function. Cartesian coordinates are also essential tools for most applied disciplines that deal with geometry, including astronomy, physics, engineering, and many more. They are the most common coordinate system used in computer graphics, computer-aided geometric design, and other geometry-related data processing.

History

The adjective *Cartesian* refers to the French mathematician and philosopher René Descartes (who used the name *Cartesius* in Latin).

The idea of this system was developed in 1637 in writings by Descartes and independently by Pierre de Fermat, although Fermat also worked in three dimensions, and did not publish the discovery. Both authors used a single axis in their treatments and have a variable length measured in reference to this axis. The concept of using a pair of axes was introduced in later work by commentators who were trying to clarify the ideas contained in Descartes' *La Géométrie*.

The development of the Cartesian coordinate system would play an intrinsic role in the development of the calculus by Isaac Newton and Gottfried Wilhelm Leibniz. Nicole Oresme, a French cleric and friend of the dauphin (later to become King Charles V) of the 14th Century, used constructions similar to Cartesian coordinates well before the time of Descartes and Fermat.

Many other coordinate systems have been developed since Descartes, such as the polar coordinates for the plane, and the spherical and cylindrical coordinates for three-dimensional space.

Definitions

Number Line

Choosing a Cartesian coordinate system for a one-dimensional space—that is, for a straight line—means choosing a point O of the line (the origin), a unit of length, and an orientation for the line. An orientation chooses which of the two half-lines determined by O is the positive, and which is negative; we then say that the line "is oriented" (or "points") from the negative half towards the positive half. Then each point p of the line can be specified by its distance from O, taken with a + or – sign depending on which half-line contains p.

A line with a chosen Cartesian system is called a number line. Every real number, whether integer, rational, or irrational, has a unique location on the line. Conversely, every point on the line can be interpreted as a number in an ordered continuum which includes the real numbers.

Cartesian Coordinates in Two Dimensions

The modern Cartesian coordinate system in two dimensions (also called a rectangular coordinate system) is defined by an ordered pair of perpendicular lines (axes), a single unit of length for both axes, and an orientation for each axis. (Early systems allowed "oblique" axes, that is, axes that did not meet at right angles.) The lines are commonly referred to as the x and y-axes where the x-axis is taken to be horizontal and the y-axis is taken to be vertical. The point where the axes meet is taken as the origin for both, thus turning each axis into a number line. For a given point P, a line is drawn through P perpendicular to the x-axis to meet it at X and second line is drawn through P perpendicular to the y-axis to meet it at Y. The coordinates of P are then X and Y interpreted as numbers x and y on the corresponding number lines. The coordinates are written as an ordered pair (x, y).

The point where the axes meet is the common origin of the two number lines and is simply called the *origin*. It is often labelled O and if so then the axes are called Ox and Oy. A plane with x and y-axes defined is often referred to as the Cartesian plane or xy plane. The value of x is called the x-coordinate or abscissa and the value of y is called the y-coordinate or ordinate.

The choices of letters come from the original convention, which is to use the latter part of the alphabet to indicate unknown values. The first part of the alphabet was used to designate known values.

In the Cartesian plane, reference is sometimes made to a unit circle or a unit hyperbola.

Cartesian Coordinates in Three Dimensions

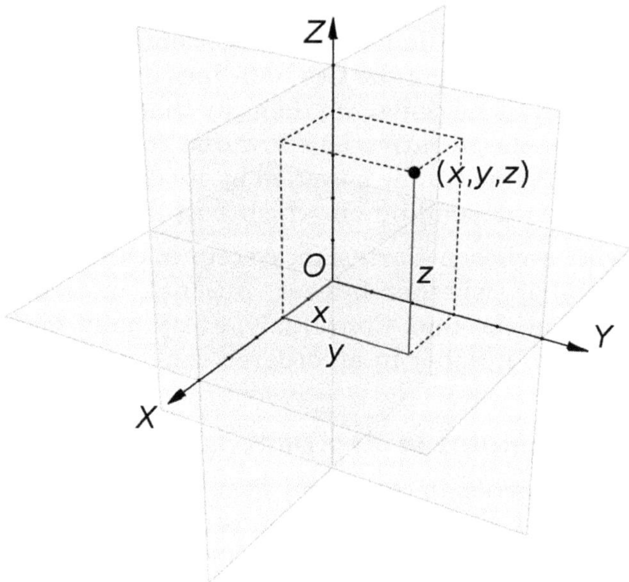

***Figure:** A three dimensional Cartesian coordinate system, with origin O and axis lines X, Y and Z, oriented as shown by the arrows. The tick marks on the axes are one length unit apart. The black dot shows the point with coordinates $X = 2$, $Y = 3$, and $Z = 4$, or (2,3,4).*

Choosing a Cartesian coordinate system for a three-dimensional space means choosing an ordered triplet of lines (axes), any two of them being perpendicular; a single unit of length for all three axes; and an orientation for each axis. As in the two-dimensional case, each axis becomes a number line. The coordinates of a point p are obtained by drawing a line through p perpendicular to each coordinate axis, and reading the points where these lines meet the axes as three numbers of these number lines.

Alternatively, the coordinates of a point p can also be taken as the (signed) distances from p to the three planes defined by the three axes. If the axes are named x, y, and z, then the x coordinate is the distance from the plane defined by the y and z axes. The distance is to be taken with the + or – sign, depending on which of the two half-spaces separated by that plane contains p. The y and z coordinates can be obtained in the same way from the (x,z) and (x,y) planes, respectively.

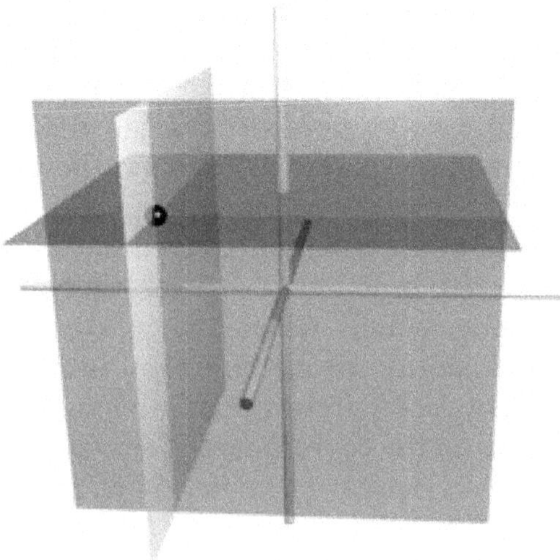

Figure: *The coordinate surfaces of the Cartesian coordinates (x, y, z). The z-axis is vertical and the x-axis is highlighted in green. Thus, the red plane shows the points with x=1, the blue plane shows the points with z=1, and the yellow plane shows the points with y=–1. The three surfaces intersect at the point P (shown as a black sphere) with the Cartesian coordinates (1, –1, 1).*

Generalizations

One can generalize the concept of Cartesian coordinates to allow axes that are not perpendicular to each other, and/or different units along each axis. In that case, each coordinate is obtained by projecting the point onto one axis along a direction that is parallel to the other axis (or, in general, to the hyper plane defined by all the other axes). In those oblique coordinate systems the computations of distances and angles is more complicated than in standard Cartesian systems, and many standard formulas (such as the Pythagorean formula for the distance) do not hold.

Notations and Conventions

The Cartesian coordinates of a point are usually written in parentheses and separated by commas, as in (10,5) or (3,5,7). The origin is often labelled with the capital letter *O*. In analytic geometry, unknown or generic coordinates are often denoted by the letters x and y on the plane, and x, y, and z in three-dimensional space. w is often used for four-dimensional space, but the rarity of such usage precludes concrete convention here. This custom comes from an old convention of algebra, to use letters near the end of the alphabet for unknown values (such as were the coordinates of points in many geometric problems), and letters near the beginning for given quantities. These conventional names are often used in other domains, such as physics and engineering. However, other letters may be used too. For example, in a graph showing how a pressure varies with time, the graph coordinates may be denoted t and P. Each axis is usually named after the coordinate which is measured along it; so one says the x-axis, the y-axis, the t-axis, etc. Another common convention for coordinate naming is to use subscripts, as in x_1, x_2, ... x_n for the n coordinates in an n-dimensional space; especially when n is greater than 3, or variable. Some authors (and many programmers) prefer the numbering x_0, x_1, ... $x_{n"1}$. These notations are especially advantageous in computer programming: by storing the coordinates of a point as an array, instead of a record, one can use iterative commands or procedure parameters instead of repeating the same commands for each coordinate.

In mathematical illustrations of two-dimensional Cartesian systems, the first coordinate (traditionally called the abscissa) is measured along a horizontal axis, oriented from left to right. The second coordinate (the ordinate) is then measured along a vertical axis, usually oriented from bottom to top. However, in computer graphics and image processing one often uses a coordinate system with the y axis pointing down (as displayed on the computer's screen). This convention developed in the 1960s (or earlier) from the way that images were originally stored in display buffers.

For three-dimensional systems, a convention is to portray the x-y plane horizontally, with the z axis added to represent height (positive up). Furthermore, there is a convention to orient "x" facing towards the viewer, biased either to the right or left. If a diagram (3D projection or 2D perspective drawing) shows the x and y axis horizontally and vertically, respectively, then the z axis should be shown pointing "out of the page" towards the viewer or camera. In such a 2D diagram of a 3D coordinate system, the z axis would appear as a line or ray pointing down and to the left or down and to the right, depending on the presumed viewer or camera perspective. In any diagram or display, the orientation of the three axes, as a whole, is arbitrary. However, the orientation of the axes relative to each other should always comply with the right-hand rule, unless specifically stated otherwise. All laws of physics and math assume this right-handedness, which ensures consistency. For 3D diagrams, the names "abscissa" and "ordinate" are rarely used for x and y, respectively. When they are, the z-coordinate is sometimes called the applicate. The words *abscissa*, *ordinate* and *applicate* are sometimes used to refer to coordinate axes rather than values.

Quadrants and Octants

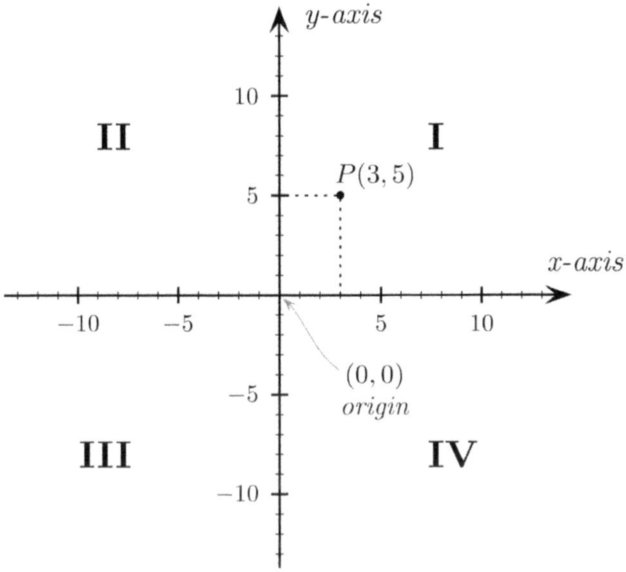

Figure: *The four quadrants of a Cartesian coordinate system.*

The axes of a two-dimensional Cartesian system divide the plane into four infinite regions, called quadrants, each bounded by two half-axes. These are often numbered from 1st to 4th and denoted by Roman numerals: I (where the signs of the two coordinates are I (+,+), II (–,+), III (–,–), and IV (+,–). When the axes are drawn according to the mathematical custom, the numbering goes counter-clockwise starting from the upper right ("northeast") quadrant. Similarly, a three-dimensional Cartesian system defines a division of space into eight regions or octants, according to the signs of the coordinates of the points. The convention used for naming a specific octant is to list its signs, e.g. (+ + +) or (– + –). The n-dimensional generalization of the quadrant and octant is the orthant, and the same naming system applies.

Cartesian Space

A Euclidean plane with a chosen Cartesian system is called a Cartesian plane. Since Cartesian coordinates are unique and non-ambiguous, the points of a Cartesian plane can be identified with all possible pairs of real numbers; that is with the Cartesian product $\mathbb{R}^2 = \mathbb{R} \times \mathbb{R}$, where $\mathbb{R}$ is the set of all reals. In the same way one defines a Cartesian space of any dimension *n*, whose points can be identified with the tuples (lists) of *n* real numbers, that is, with $\mathbb{R}^n$.

Chapter

5

Game Theory

Game theory is a study of strategic decision making. More formally, it is "the study of mathematical models of conflict and cooperation between intelligent rational decision-makers." An alternative term suggested "as a more descriptive name for the discipline" is *interactive decision theory*. Game theory is mainly used in economics, political science, and psychology, as well as logic and biology. The subject first addressed zero-sum games, such that one person's gains exactly equal net losses of the other participant(s). Today, however, game theory applies to a wide range of class relations, and has developed into an umbrella term for the logical side of science, to include both human and non-humans, like computers. Classic uses include a sense of balance in numerous games, where each person has found or developed a tactic that cannot successfully better his results, given the other approach.

Modern game theory began with the idea regarding the existence of mixed-strategy equilibria in two-person zero-sum games and its proof by John von Neumann. Von Neumann's original proof used Brouwer's fixed-point theorem on continuous mappings into compact convex sets, which became a standard method in game theory and mathematical economics. His paper was followed by his 1944 book *Theory of Games and Economic Behaviour*, with Oskar Morgenstern, which considered cooperative games of several players. The second edition of this book provided an axiomatic theory of expected utility, which allowed mathematical statisticians and economists to treat decision-making under uncertainty.

This theory was developed extensively in the 1950s by many scholars. Game theory was later explicitly applied to biology in the 1970s, although similar developments go back at least as far as the 1930s. Game theory has been widely recognized as an important tool in many fields. Eight game-theorists have won the Nobel Memorial Prize in Economic Sciences, and John Maynard Smith was awarded the Crafoord Prize for his application of game theory to biology.

Representation of Games

The games studied in game theory are well-defined mathematical objects. A game consists of a set of players, a set of moves (or strategies) available to those players, and a specification of payoffs for each combination of strategies. Most cooperative games are presented in the characteristic function form, while the extensive and the normal forms are used to define noncooperative games.

Extensive Form

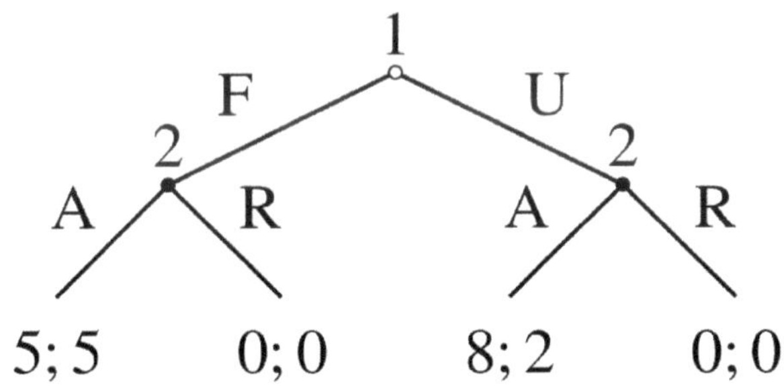

Figure: *An extensive form game*

The extensive form can be used to formalize games with a time sequencing of moves. Games here are played on trees (as pictured to the left). Here each vertex (or node) represents a point of choice for a player. The player is specified by a number listed by the vertex. The lines out of the vertex represent a possible action for that player. The payoffs are specified at the bottom of the tree. The extensive form can be viewed as a multi-player generalization of a decision tree. (Fudenberg & Tirole 1991, p. 67)

In the game pictured to the left, there are two players. *Player 1* moves first and chooses either *F* or *U*. *Player 2* sees *Player 1*'s move and then chooses *A* or *R*. Suppose that *Player 1* chooses *U* and then *Player 2* chooses *A*, then *Player 1* gets 8 and *Player 2* gets 2.

The extensive form can also capture simultaneous-move games and games with imperfect information. To represent it, either a dotted line connects different vertices to represent them as being part of the same information set (i.e., the players do not know at which point they are), or a closed line is drawn around them.

Characteristic Function Form

In games that possess removable utility separate rewards are not given; rather, the characteristic function decides the payoff of each unity. The idea is that the unity that is 'empty', so to speak, does not receive a reward at all.

The origin of this form is to be found in John von Neumann and Oskar Morgenstern's book; when looking at these instances, they guessed that when a union C appears, it works against the fraction $\left(\frac{N}{C}\right)$ as if two individuals were playing a normal game. The balanced payoff of C is a basic function. Although there are differing examples that help determine coalitional amounts from normal games, not all appear that in their function form can be derived from such.

Formally, a characteristic function is seen as: (N,v), where N represents the group of people and $v: 2^N \rightarrow R$ is a normal utility. Such characteristic functions have expanded to describe games where there is no removable utility.

Partition Function Form

The characteristic function form ignores the possible externalities of coalition formation. In the partition function form the payoff of a coalition depends not only on its members, but also on the way the rest of the players are partitioned (Thrall & Lucas 1963).

General and Applied Uses

As a method of applied mathematics, game theory has been used to study a wide variety of human and animal behaviours. It was initially developed in economics to understand a large collection of economic behaviours, including behaviours of firms, markets, and consumers. The use of game theory in the social sciences has expanded, and game theory has been applied to political, sociological, and psychological behaviours as well.

Game-theoretic analysis was initially used to study animal behaviour by Ronald Fisher in the 1930s (although even Charles Darwin makes a few informal game-theoretic statements). This work predates the name "game theory", but it shares many important features with this field. The developments in economics were later applied to biology largely by John Maynard Smith in his book *Evolution and the Theory of Games.*

In addition to being used to describe, predict, and explain behaviour, game theory has also been used to develop theories of ethical or normative behaviour and to prescribe such behaviour. In economics and philosophy, scholars have applied game theory to help in the understanding of good or proper behaviour. Game-theoretic arguments of this type can be found as far back as Plato.

Description and Modelling

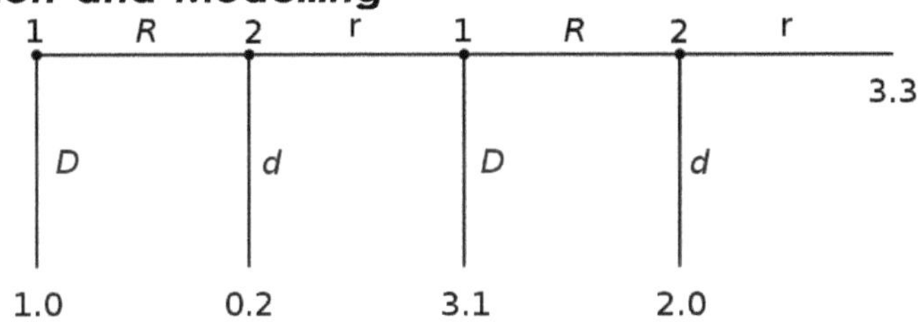

Figure: *A three stage centipede game*

The first known use is to describe and model how human populations behave. Some scholars believe that by finding the equilibria of games they can predict how actual human populations will behave when confronted with situations analogous to the game being studied. This particular view of game theory has come under recent criticism. First, it is criticized because the assumptions made by game theorists are often violated. Game theorists may assume players always act in a way to directly maximize their wins (the Homo economicus model), but in practice, human behaviour often deviates from this model. Explanations of this phenomenon are many; irrationality, new models of deliberation, or even different motives (like that of altruism). Game theorists respond by comparing their assumptions to those used in physics. Thus while their assumptions do not always hold, they can treat game theory as a reasonable scientific ideal akin to the models used by physicists. However, in the centipede game, guess 2/3 of the average game, and the dictator game, people regularly do not play Nash equilibria. These experiments have demonstrated that individuals do not play equilibrium strategies. There is an ongoing debate regarding the importance of these experiments.

Alternatively, some authors claim that Nash equilibria do not provide predictions for human populations, but rather provide an explanation for why populations that play Nash equilibria remain in that state. However, the question of how populations reach those points remains open.

Some game theorists have turned to evolutionary game theory in order to resolve these issues. These models presume either no rationality or bounded rationality on the part of players. Despite the name, evolutionary game theory does not necessarily presume natural selection in the biological sense. Evolutionary game theory includes both biological as well as cultural evolution and also models of individual learning (for example, fictitious play dynamics).

Prescriptive or Normative Analysis

Table: *The Prisoner's Dilemma*

	Cooperate	***Defect***
Cooperate	-1, -1	-10, 0
Defect	0, -10	-5, -5

On the other hand, some scholars see game theory not as a predictive tool for the behaviour of human beings, but as a suggestion for how people ought to behave. Since a strategy, corresponding to a Nash equilibrium of a game constitutes one's best response to the actions of the other players – provided they are in (the same) Nash equilibrium –, playing a strategy that is part of a Nash equilibrium seems appropriate. However, the rationality of such a decision has been proved only for special cases.

This normative use of game theory has also come under criticism. First, in some cases it is appropriate to play a non-equilibrium strategy if one expects others to play non-equilibrium strategies as well.

Second, the prisoner's dilemma presents another potential counterexample. In the prisoner's dilemma, each player pursuing his own self-interest leads both players to be worse off than had they not pursued their own self-interests.

Economics and Business

Game theory is a major method used in mathematical economics and business for modelling competing behaviours of interacting agents. Applications include a wide array of economic phenomena and approaches, such as auctions, bargaining, mergers & acquisitions pricing, fair division, duopolies, oligopolies, social network formation, agent-based computational economics, general equilibrium, mechanism design, and voting systems, and across such broad areas as experimental economics, behavioural economics, information economics, industrial organization, and political economy.

This research usually focuses on particular sets of strategies known as equilibria in games. These "solution concepts" are usually based on what is required by norms of rationality. In non-cooperative games, the most famous of these is the Nash equilibrium. A set of strategies is a Nash equilibrium if each represents a best response to the other strategies. So, if all the players are playing the strategies in a Nash equilibrium, they have no unilateral incentive to deviate, since their strategy is the best they can do given what others are doing.

The payoffs of the game are generally taken to represent the utility of individual players. Often in modelling situations the payoffs represent money, which presumably corresponds to an individual's utility. This assumption, however, can be faulty.

A prototypical paper on game theory in economics begins by presenting a game that is an abstraction of a particular economic situation. One or more solution concepts are chosen, and the author demonstrates which strategy sets in the presented game are equilibria of the appropriate type. Naturally one might wonder to what use should this information be put. Economists and business professors suggest two primary uses (noted above): *descriptive* and *prescriptive*.

Political Science

The application of game theory to political science is focused in the overlapping areas of fair division, political economy, public choice, war bargaining, positive political theory, and social choice theory. In each of these areas, researchers have developed game-theoretic models in which the players are often voters, states, special interest groups, and politicians.

For early examples of game theory applied to political science. In his book *An Economic Theory of Democracy* (Downs 1957), he applies the Hotelling firm location model to the political process. In the Downsian model, political candidates commit to ideologies on a one-dimensional policy space. Downs first shows how the political candidates will converge to the ideology preferred by the median voter if voters are fully informed, but then argues that voters choose to remain rationally ignorant which allows for candidate divergence.

A game-theoretic explanation for democratic peace is that public and open debate in democracies send clear and reliable information regarding their intentions to other states. In contrast, it is difficult to know the intentions of nondemocratic leaders, what effect concessions will have, and if promises will be kept. Thus there will be mistrust and unwillingness to make concessions if at least one of the parties in a dispute is a non-democracy (Levy &Razin 2003).

Biology

Table: *The hawk-dove game*

	Hawk	*Dove*
Hawk	20, 20	80, 40
Dove	40, 80	60, 60

The payoffs for games in biology, unlike those in economics, are often interpreted as corresponding to fitness. In addition, the focus has been less on equilibria that correspond to a notion of rationality and more on ones that would be maintained by evolutionary forces. The best known equilibrium in biology is known as the *evolutionarily stable strategy* (or ESS), and was first introduced in (Smith & Price 1973). Although its initial motivation did not involve any of the mental requirements of the Nash equilibrium, every ESS is a Nash equilibrium.

In biology, game theory has been used to understand many different phenomena. It was first used to explain the evolution (and stability) of the approximate 1:1 sex ratios. (Fisher 1930) suggested that the 1:1 sex ratios are a result of evolutionary forces acting on individuals who could be seen as trying to maximize their number of grandchildren.

Additionally, biologists have used evolutionary game theory and the ESS to explain the emergence of animal communication (Harper & Maynard Smith 2003). The analysis of signalling games and other communication games has provided insight into the evolution of communication among animals. For example, the mobbing behaviour of many species, in which a large number of prey animals attack a larger predator, seems to be an example of spontaneous emergent organization. Ants have also been shown to exhibit feed-forward behaviour akin to fashion.

Biologists have used the game of chicken to analyze fighting behaviour and territoriality.

Maynard Smith, in the preface to *Evolution and the Theory of Games*, writes, "paradoxically, it has turned out that game theory is more readily applied to biology than to the field of economic behaviour for which it was originally designed". Evolutionary game theory has been used to explain many seemingly incongruous phenomena in nature.

One such phenomenon is known as biological altruism. This is a situation in which an organism appears to act in a way that benefits other organisms and is detrimental to itself. This is distinct from traditional notions of altruism because such actions are not conscious, but appear to be evolutionary adaptations to increase overall fitness. Examples can be found in species ranging from vampire bats that regurgitate blood they have obtained from a night's hunting and give it to group members who have failed to feed, to worker bees that care for the queen bee for their entire lives and never mate, to Vervet monkeys that warn group members of a predator's approach, even when it endangers that individual's chance of survival. All of these actions increase the overall fitness of a group, but occur at a cost to the individual.

Evolutionary game theory explains this altruism with the idea of kin selection. Altruists discriminate between the individuals they help and favour relatives. Hamilton's rule explains the evolutionary reasoning behind this selection with the equation $c<b*r$ where the cost (c) to the altruist must be less than the benefit (b) to the recipient multiplied by the coefficient of relatedness (r). The more closely related two organisms are causes the incidences of altruism to increase because they share many of the same alleles. This means that the altruistic individual, by ensuring that the alleles of its close relative are passed on, (through survival of its offspring) can forgo the option of having offspring itself because the same number of alleles are passed on. Helping a sibling for example (in diploid animals), has a coefficient of ½, because (on average) an individual shares ½ of the alleles in its sibling's offspring. Ensuring that enough of a sibling's offspring survive to adulthood precludes the necessity of the altruistic individual producing offspring. The coefficient values depend heavily on the scope of the playing field; for example if the choice of whom to favour includes all genetic living things, not just all relatives, we assume the discrepancy between all humans only accounts for approximately 1% of the diversity in the playing field, a co-efficient that was ½ in the smaller field becomes 0.995. Similarly if it is considered that information other than that of a genetic nature (e.g. epigenetics, religion, science, etc.) persisted through time the playing field becomes larger still, and the discrepancies smaller.

Computer Science and Logic

Game theory has come to play an increasingly important role in logic and in computer science. Several logical theories have a basis in game semantics. In addition, computer scientists have used games to model interactive computations. Also, game theory provides a theoretical basis to the field of multi-agent systems.

Separately, game theory has played a role in online algorithms. In particular, the *k*-server problem, which has in the past been referred to as *games with moving costs* and *request-answer games* (Ben David, Borodin & Karp et al. 1994).

Yao's principle is a game-theoretic technique for proving lower bounds on the computational complexity of randomized algorithms, and especially of online algorithms.

The emergence of the internet has motivated the development of algorithms for finding equilibria in games, markets, computational auctions, peer-to-peer systems, and security and information markets. Algorithmic game theory and within it algorithmic mechanism design combine computational algorithm design and analysis of complex systems with economic theory.

Philosophy

Table: *Stag hunt*

	Stag	***Hare***
Stag	3, 3	0, 2
Hare	2, 0	2, 2

Game theory has been put to several uses in philosophy. Responding to two papers by W.V.O. Quine (1960, 1967), Lewis (1969) used game theory to develop a philosophical account of convention. In so doing, he provided the first analysis of common knowledge and employed it in analyzing play in coordination games. In addition, he first suggested that one can understand meaning in terms of signalling games.

This later suggestion has been pursued by several philosophers since Lewis (Skyrms (1996), Grim, Kokalis, and Alai-Tafti et al. (2004)). Following Lewis (1969) game-theoretic account of conventions, Edna Ullmann-Margalit (1977) and Bicchieri (2006) have developed theories of social norms that define them as Nash equilibria that result from transforming a mixed-motive game into a coordination game.

Game theory has also challenged philosophers to think in terms of interactive epistemology: what it means for a collective to have common beliefs or knowledge, and what are the consequences of this knowledge for the social outcomes resulting from agents' interactions. Philosophers who have worked in this area include Bicchieri (1989, 1993), Skyrms (1990), and Stalnaker (1999).

In ethics, some authors have attempted to pursue the project, begun by Thomas Hobbes, of deriving morality from self-interest. Since games like the prisoner's dilemma present an apparent conflict between morality and self-interest, explaining why cooperation is required by self-interest is an important component of this project. This general strategy is a component of the general social contract view in political philosophy.

Other authors have attempted to use evolutionary game theory in order to explain the emergence of human attitudes about morality and corresponding animal behaviours. These authors look at several games including the prisoner's dilemma, stag hunt, and the Nash bargaining game as providing an explanation for the emergence of attitudes about morality.

Some assumptions used in some parts of game theory have been challenged in philosophy; for example, psychological egoism states that rationality reduces to self-interest—a claim debated among philosophers.

Types of Games

Cooperative or Non-cooperative

A game is *cooperative* if the players are able to form binding commitments. For instance the legal system requires them to adhere to their promises. In noncooperative games this is not possible. Often it is assumed that *communication* among players is allowed in cooperative games, but not in noncooperative ones. However, this classification on two binary criteria has been questioned, and sometimes rejected (Harsanyi 1974).

Of the two types of games, noncooperative games are able to model situations to the finest details, producing accurate results. Cooperative games focus on the game at large.

Considerable efforts have been made to link the two approaches. The so-called Nash-programme has already established many of the cooperative solutions as noncooperativeequilibria.

Hybrid games contain cooperative and non-cooperative elements. For instance, coalitions of players are formed in a cooperative game, but these play in a non-cooperative fashion.

Symmetric and Asymmetric

***Table:** An asymmetric game*

	E	*F*
E	1, 2	0, 0
F	0, 0	1, 2

A symmetric game is a game where the payoffs for playing a particular strategy depend only on the other strategies employed, not on who is playing them. If the identities of the players can be changed without changing the payoff to the strategies, then a game is symmetric. Many of the commonly studied 2×2 games are symmetric. The standard representations of chicken, the prisoner's dilemma, and the stag hunt are all symmetric games. Some scholars would consider certain asymmetric games as examples of these games as well. However, the most common payoffs for each of these games are symmetric.

Most commonly studied asymmetric games are games where there are not identical strategy sets for both players. For instance, the ultimatum game and similarly the dictator game have different strategies for each player. It is possible, however, for a game to have identical strategies for both players, yet be asymmetric. For example, the game pictured to the right is asymmetric despite having identical strategy sets for both players.

Zero-sum and non-zero-sum

***Table:** A zero-sum game*

	A	*B*
A	–1, 1	3, –3
B	0, 0	–2, 2

Zero-sum games are a special case of constant-sum games, in which choices by players can neither increase nor decrease the available resources. In zero-sum games the total benefit to all players in the game, for every combination of strategies, always adds to zero (more informally, a player benefits only at the equal expense of others). Poker exemplifies a zero-sum game (ignoring the possibility of the house's cut), because one wins exactly the amount one's opponents lose. Other zero-sum games include matching pennies and most classical board games including Go and chess.

Many games studied by game theorists (including the infamous prisoner's dilemma) are non-zero-sum games, because the outcome has net results greater or less than zero. Informally, in non-zero-sum games, a gain by one player does not necessarily correspond with a loss by another. Constant-sum games correspond to activities like theft and gambling, but not to the fundamental economic situation in which there are potential gains from trade. It is possible to transform any game into a (possibly asymmetric) zero-sum game by adding an additional dummy player (often called "the board"), whose losses compensate the players' net winnings.

Simultaneous and Sequential

Simultaneous games are games where both players move simultaneously, or if they do not move simultaneously, the later players are unaware of the earlier players' actions (making them *effectively* simultaneous). Sequential games (or dynamic games) are games where later players have some knowledge about earlier actions. This need not be perfect information about every action of earlier players; it might be very little knowledge. For instance, a player may know that an earlier player did not perform one particular action, while he does not know which of the other available actions the first player actually performed.

The difference between simultaneous and sequential games is captured in the different representations discussed above. Often, normal form is used to represent simultaneous games, and extensive form is used to represent sequential ones. The transformation of extensive to normal form is one way, meaning that multiple extensive form games correspond to the same normal form. Consequently, notions of equilibrium for simultaneous games are insufficient for reasoning about sequential games.

In short, the differences between sequential and simultaneous games are as follows:

	Sequential	***Simultaneous***
Normally denoted by:	Decision Trees	Payoff Matrices
Prior knowledge of opponent's move:	Yes	No
Time Axis:	Yes	No
Also known as:	Extensive Game	Strategic Game

Perfect Information and Imperfect Information

An important subset of sequential games consists of games of perfect information. A game is one of perfect information if all players know the moves previously made by all other players. Thus, only sequential games can be games of perfect information because players in simultaneous games do not know the actions of the other players. Most games studied in game theory are imperfect-information games. Interesting examples of perfect-information games include the ultimatum game and centipede game. Recreational games of perfect information games include chess, go, and mancala. Many card games are games of imperfect information, for instance poker or contract bridge.

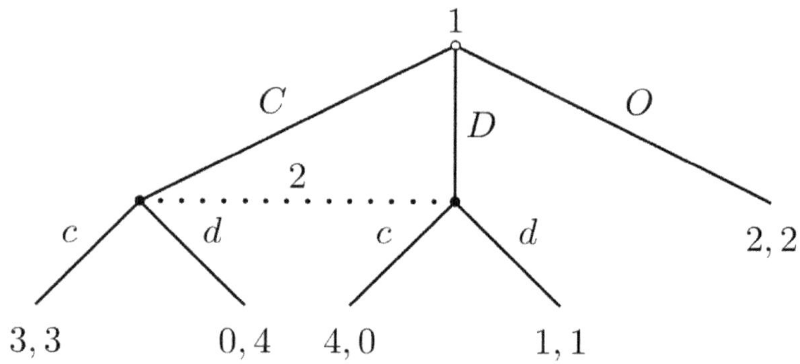

Figure: *A game of imperfect information (the dotted line represents ignorance on the part of player 2, formally called an information set)*

Perfect information is often confused with complete information, which is a similar concept. Complete information requires that every player know the strategies and payoffs available to the other players but not necessarily the actions taken. Games of incomplete information can be reduced, however, to games of imperfect information by introducing "moves by nature" (Leyton-Brown &Shoham 2008, p. 60).

Combinatorial Games

Games in which the difficulty of finding an optimal strategy stems from the multiplicity of possible moves are called combinatorial games. Examples include chess and go. Games that involve imperfect or incomplete information may also have a strong combinatorial character, for instance backgammon. There is no unified theory addressing combinatorial elements in games. There are, however, mathematical tools that can solve particular problems and answer general questions.

Games of perfect information have been studied in combinatorial game theory, which has developed novel representations, e.g. surreal numbers, as well as combinatorial and algebraic (and sometimes non-constructive) proof methods to solve games of certain types, including "loopy" games that may result in infinitely long sequences of moves. These methods address games with higher combinatorial complexity than those usually considered in traditional (or "economic") game theory. A typical game that has been solved this way is hex. A related field of study, drawing from computational complexity theory, is game complexity, which is concerned with estimating the computational difficulty of finding optimal strategies.

Research in artificial intelligence has addressed both perfect and imperfect (or incomplete) information games that have very complex combinatorial structures (like chess, go, or backgammon) for which no provable optimal strategies have been found. The practical solutions involve computational heuristics, like alpha-beta pruning or use of artificial neural networks trained by reinforcement learning, which make games more tractable in computing practice.

Infinitely Long Games

Games, as studied by economists and real-world game players, are generally finished in finitely many moves. Pure mathematicians are not so constrained, and set theorists in particular study games that last for infinitely many moves, with the winner (or other payoff) not known until *after* all those moves are completed.

The focus of attention is usually not so much on what is the best way to play such a game, but simply on whether one or the other player has a winning strategy. (It can be proven, using the axiom of choice, that there are games—even with perfect information, and where the only outcomes are "win" or "lose"—for which *neither* player has a winning strategy.) The existence of such strategies, for cleverly designed games, has important consequences in descriptive set theory.

Discrete and Continuous Games

Much of game theory is concerned with finite, discrete games, that have a finite number of players, moves, events, outcomes, etc. Many concepts can be extended, however. Continuous games allow players to choose a strategy from a continuous strategy set. For instance, Cournot competition is typically modelled with players' strategies being any non-negative quantities, including fractional quantities.

Differential Games

Differential games such as the continuous pursuit and evasion game are continuous games where the evolution of the players' state variables is governed by differential equations. The problem of finding an optimal strategy in a differential game is closely related to the optimal control theory. In particular, there are two types of strategies: the open-loop strategies are found using the Pontryagin maximum principle while the closed-loop strategies are found using Bellman's Dynamic Programming method.

A particular case of differential games are the games with random time horizon. In such games, the terminal time is a random variable with a given probability distribution function. Therefore, the players maximize the mathematical expectation of the cost function. It was shown that the modified optimization problem can be reformulated as a discounted differential game over an infinite time interval.

Many-player and Population Games

Games with an arbitrary, but finite, number of players are often called n-person games (Luce &Raiffa 1957). Evolutionary game theory considers games involving a population of decision makers, where the frequency with which a particular decision is made can change over time in response to the decisions made by all individuals in the population. In biology, this is intended to model (biological) evolution, where genetically programmed organisms pass along some of their strategy programming to their offspring. In economics, the same theory is intended to capture population changes because people play the game many times within their lifetime, and consciously (and perhaps rationally) switch strategies (Webb 2007).

Stochastic Outcomes (and Relation to other Fields)

Individual decision problems with stochastic outcomes are sometimes considered "one-player games". These situations are not considered game theoretical by some authors. They may be modelled using similar tools within the related disciplines of decision theory, operations research, and areas of artificial intelligence, particularly AI planning (with uncertainty) and multi-agent system. Although these fields may have different motivators, the mathematics involved are substantially the same, e.g. using Markov decision processes (MDP).

Stochastic outcomes can also be modelled in terms of game theory by adding a randomly acting player who makes "chance moves", also known as "moves by nature" (Osborne & Rubinstein 1994). This player is not typically considered a third player in what is otherwise a two-player game, but merely serves to provide a roll of the dice where required by the game.

For some problems, different approaches to modelling stochastic outcomes may lead to different solutions. For example, the difference in approach between MDPs and the minimax solution is that the latter considers the worst-case over a set of adversarial moves, rather than reasoning in expectation about these moves given a fixed probability distribution. The minimax approach may be advantageous where stochastic models of uncertainty are not available, but may also be overestimating extremely unlikely (but costly) events, dramatically swaying the strategy in such scenarios if it is assumed that an adversary can force such an event to happen.

General models that include all elements of stochastic outcomes, adversaries, and partial or noisy observability (of moves by other players) have also been studied. The "gold standard" is considered to be partially observable stochastic game (POSG), but few realistic problems are computationally feasible in POSG representation.

Metagames

These are games the play of which is the development of the rules for another game, the target or subject game. Metagames seek to maximize the utility value of the rule set developed. The theory of metagames is related to mechanism design theory.

The term metagame analysis is also used to refer to a practical approach developed by Nigel Howard (Howard 1971) whereby a situation is framed as a strategic game in which stakeholders try to realise their objectives by means of the options available to them. Subsequent developments have led to the formulation of confrontation analysis.

History

Early discussions of examples of two-person games occurred long before the rise of modern, mathematical game theory. The first known discussion of game theory occurred in a letter written by James Waldegrave in 1713. In this letter, Waldegrave provides a minimaxmixed strategy solution to a two-person version of the card game le Her. James Madison made what we now recognize as a game-theoretic analysis of the ways states can be expected to behave under different systems of taxation. In his 1838 *Recherchessur les principesmathématiques de la théorie des richesses* (*Researches into the Mathematical Principles of the Theory of Wealth*), Antoine AugustinCournot considered a duopoly and presents a solution that is a restricted version of the Nash equilibrium.

The Danish mathematician Zeuthen proved that the mathematical model had a winning strategy by using Brouwer's fixed point theorem. In his 1938 book *Applications aux Jeux de Hasard* and earlier notes, Émile Borel proved a minimax theorem for two-person zero-sum matrix games only when the pay-off matrix was symmetric. Borel conjectured that non-existence of mixed-strategy equilibria in two-person zero-sum games would occur, a conjecture that was proved false.

Game theory did not really exist as a unique field until John von Neumann published a paper in 1928. Von Neumann's original proof used Brouwer's fixed-point theorem on continuous mappings into compact convex sets, which became a standard method in game theory and mathematical economics. His paper was followed by his 1944 book *Theory of Games and Economic Behaviour*. The second edition of this book provided an axiomatic theory of utility, which reincarnated Daniel Bernoulli's old theory of utility (of the money) as an independent discipline. Von Neumann's work in game theory culminated in this 1944 book. This foundational work contains the method for finding mutually consistent solutions for two-person zero-sum games. During the following time period, work on game theory was primarily focused on cooperative game theory, which analyzes optimal strategies for groups of individuals, presuming that they can enforce agreements between them about proper strategies.

In 1950, the first mathematical discussion of the prisoner's dilemma appeared, and an experiment was undertaken by notable mathematicians Merrill M. Flood and Melvin Dresher, as part of the RAND corporation's investigations into game theory. Rand pursued the studies because of possible applications to global nuclear strategy. Around this same time, John Nash developed a criterion for mutual consistency of players' strategies, known as Nash equilibrium, applicable to a wider variety of games than the criterion proposed by von Neumann and Morgenstern. This equilibrium is sufficiently general to allow for the analysis of non-cooperative games in addition to cooperative ones. Game theory experienced a flurry of activity in the 1950s, during which time the concepts of the core, the extensive form game, fictitious play, repeated games, and the Shapley value were developed. In addition, the first applications of game theory to philosophy and political science occurred during this time.

In 1965, Reinhard Selten introduced his solution concept of subgame perfect equilibria, which further refined the Nash equilibrium (later he would introduce trembling hand perfection as well). In 1967,

John Harsanyi developed the concepts of complete information and Bayesian games. Nash, Selten and Harsanyi became Economics Nobel Laureates in 1994 for their contributions to economic game theory.

In the 1970s, game theory was extensively applied in biology, largely as a result of the work of John Maynard Smith and his evolutionarily stable strategy. In addition, the concepts of correlated equilibrium, trembling hand perfection, and common knowledge were introduced and analyzed.

In 2005, game theorists Thomas Schelling and Robert Aumann followed Nash, Selten and Harsanyi as Nobel Laureates. Schelling worked on dynamic models, early examples of evolutionary game theory. Aumann contributed more to the equilibrium school, introducing an equilibrium coarsening, correlated equilibrium, and developing an extensive formal analysis of the assumption of common knowledge and of its consequences.

In 2007, Leonid Hurwicz, together with Eric Maskin and Roger Myerson, was awarded the Nobel Prize in Economics "for having laid the foundations of mechanism design theory." Myerson's contributions include the notion of proper equilibrium, and an important graduate text: *Game Theory, Analysis of Conflict* (Myerson 1997). Hurwicz introduced and formalized the concept of incentive compatibility.

Interest

Interest is a fee paid by a borrower of assets to the owner as a form of compensation for the use of the assets. It is most commonly the price paid for the use of borrowed money, or money earned by deposited funds.

When money is borrowed, interest is typically paid to the lender as a percentage of the principal, the amount owed to the lender. The percentage of the principal that is paid as a fee over a certain period of time (typically one month or year) is called the interest rate. A bank deposit will earn interest because the bank is paying for the use of the deposited funds. Assets that are sometimes lent with interest include money, shares, consumer goods through hire purchase, major assets such as aircraft, and even entire factories in finance lease arrangements. The interest is calculated upon the value of the assets in the same manner as upon money.

Interest is compensation to the lender, for a) risk of principal loss, called credit risk; and b) forgoing other investments that could have been made with the loaned asset. These forgone investments are known as the opportunity cost. Instead of the lender using the assets directly, they are advanced to the borrower. The borrower then enjoys the benefit of using the assets ahead of the effort required to pay for them, while the lender enjoys the benefit of the fee paid by the borrower for the privilege. In economics, interest is considered the price of credit.

Interest is often compounded, which means that interest is earned on prior interest in addition to the principal. The total amount of debt grows exponentially, most notably when compounded at infinitesimally small intervals, and its mathematical study led to the discovery of the number *e*. However, in practice, interest is most often calculated on a daily, monthly, or yearly basis, and its impact is influenced greatly by its compounding rate.

History of Interest

According to historian Paul Johnson, the lending of "food money" was commonplace in Middle East civilizations as far back as 5000BC. They regarded interest as legitimate since acquired seeds and animals could "reproduce themselves"; whilst the ancient Jewish religious prohibitions against usury were a "different view".

In the Roman Empire interest rates were usually calculated on a monthly basis and set as multiples of 12, apparently for expedient calculation by the wealthy private individuals that did most of the moneylending.

The First Council of Nicaea, in 325, forbade clergy from engaging in usury which was defined as lending on interest above 1 percent per month (12.7% APR). Later ecumenical councils applied this regulation to the laity. Catholic Church opposition to interest hardened in the era of scholastics, when even defending it was considered a heresy. St. Thomas Aquinas, the leading theologian of the Catholic Church, argued that the charging of interest is wrong because it amounts to "double charging", charging for both the thing and the use of the thing.

In the medieval economy, loans were entirely a consequence of necessity (bad harvests, fire in a workplace) and, under those conditions, it was considered morally reproachable to charge interest. It was also considered morally dubious, since no goods were produced through the lending of money, and thus it should not be compensated, unlike other activities with direct physical output such as blacksmithing or farming. For the same reason, interest has often been looked down upon in

Islamic civilization, with most scholars agreeing that the Qur'an explicitly forbids charging interest.

Medieval jurists developed several financial instruments to encourage responsible lending and circumvent prohibitions on usury, such as the Contractumtrinius.

In the Renaissance era, greater mobility of people facilitated an increase in commerce and the appearance of appropriate conditions for entrepreneurs to start new, lucrative businesses. Given that borrowed money was no longer strictly for consumption but for production as well, interest was no longer viewed in the same manner. The School of Salamanca elaborated on various reasons that justified the charging of interest: the person who received a loan benefited, and one could consider interest as a premium paid for the risk taken by the loaning party.

There was also the question of opportunity cost, in that the loaning party lost other possibilities of using the loaned money. Finally and perhaps most originally was the consideration of money itself as merchandise, and the use of one's money as something for which one should receive a benefit in the form of interest. Martín de Azpilcueta also considered the effect of time. Other things being equal, one would prefer to receive a given good now rather than in the future. This preference indicates greater value. Interest, under this theory, is the payment for the time the loaning individual is deprived of the money.

Economically, the interest rate is the cost of capital and is subject to the laws of supply and demand of the money supply. The first attempt to control interest rates through manipulation of the money supply was made by the French Central Bank in 1847.

The first formal studies of interest rates and their impact on society were conducted by Adam Smith, Jeremy Bentham and Mirabeau during the birth of classic economic thought. In the late 19th century leading Swedish economist Knut Wicksell in his 1898 *Interest and Prices* elaborated a comprehensive theory of economic crises based upon a distinction between natural and nominal interest rates. In the early 20th century, Irving Fisher made a major breakthrough in the economic analysis of interest rates by distinguishing nominal interest from real interest. Several perspectives on the nature and impact of interest rates have arisen since then.

The latter half of the 20th century saw the rise of interest-free Islamic banking and finance, a movement that attempts to apply religious law developed in the medieval period to the modern economy.

Some entire countries, including Iran, Sudan, and Pakistan, have taken steps to eradicate interest from their financial systems altogether. Rather than charging interest, the interest-free lender shares the risk by investing as a partner in profit loss sharing scheme, because predetermined loan repayment as interest is prohibited, as well as making money out of money is unacceptable.

All financial transactions must be asset-backed and it does not charge any "fee" for the service of lending.

Types of Interest

Simple Interest: Simple interest is calculated only on the principal amount, or on that portion of the principal amount that remains unpaid.

The amount of simple interest is calculated according to the following formula:

$$I_{simp} = r \cdot B_0 \cdot m_t$$

where r is the period interest rate (I/m), B_0 the initial balance and m_t the number of time periods elapsed.

To calculate the period interest rate r, one divides the interest rate I by the number of periods m_t.

For example, imagine that a credit card holder has an outstanding balance of \$2500 and that the simple interest rate is 12.99% per annum. The interest added at the end of 3 months would be,

$$I_{simp} = \left(\frac{0.1299}{12} \cdot \$2500\right) \cdot 3 = \$81.19$$

and they would have to pay \$2581.19 to pay off the balance at this point.

If instead they make interest-only payments for each of those 3 months at the period rate r, the amount of interest paid would be,

$$I = \left(\frac{0.1299}{12} \cdot \$2500\right) \cdot 3 = (\$27.0625 / month) \cdot 3 = \$81.19$$

Their balance at the end of 3 months would still be \$2500.

In this case, the time value of money is not factored in. The steady payments have an additional cost that needs to be considered when comparing loans. For example, given a \$100 principal:

- Credit card debt where \$1/day is charged: 1/100 = 1%/day = 7%/week = 365%/year.

- Corporate bond where the first \$3 are due after six months, and the second \$3 are due at the year's end: (3+3)/100 = 6%/year.
- Certificate of deposit (GIC) where \$6 is paid at the year's end: 6/100 = 6%/year.

There are two complications involved when comparing different simple interest bearing offers.

1. When rates are the same but the periods are different a direct comparison is inaccurate because of the time value of money. Paying \$3 every six months costs more than \$6 paid at year end so, the 6% bond cannot be 'equated' to the 6% GIC.
2. When interest is due, but not paid, does it remain 'interest payable', like the bond's \$3 payment after six months or, will it be added to the balance due? In the latter case it is no longer simple interest, but compound interest.

A bank account that offers only simple interest, that money can freely be withdrawn from is unlikely, since withdrawing money and immediately depositing it again would be advantageous.

Composition of Interest Rates

In economics, interest is considered the price of credit, therefore, it is also subject to distortions due to inflation. The nominal interest rate, which refers to the price before adjustment to inflation, is the one visible to the consumer (*i.e.*, the interest tagged in a loan contract, credit card statement, etc.). Nominal interest is composed of the real interest rate plus inflation, among other factors. A simple formula for the nominal interest is:

$$i = r + \pi$$

Where i is the nominal interest, r is the real interest and π is inflation.

This formula attempts to measure the value of the interest in units of stable purchasing power. However, if this statement were true, it would imply at least two misconceptions. First, that all interest rates within an area that shares the same inflation (that is, the same country) should be the same. Second, that the lenders know the inflation for the period of time that they are going to lend the money.

One reason behind the difference between the interest that yields a treasury bond and the interest that yields a mortgage loan is the risk that the lender takes from lending money to an economic agent.

In this particular case, a government is more likely to pay than a private citizen. Therefore, the interest rate charged to a private citizen is larger than the rate charged to the government.

To take into account the information asymmetry aforementioned, both the value of inflation and the real price of money are changed to their expected values resulting in the following equation:

$$i_t = r_{(t+1)} + \pi_{(t+1)} + \sigma$$

Here, i_t is the nominal interest at the time of the loan, $r_{(t+1)}$ is the real interest expected over the period of the loan, $\pi_{(t+1)}$ is the inflation expected over the period of the loan and σ is the representative value for the risk engaged in the operation.

Cumulative Interest or Return

The calculation for cumulative interest is (FV/PV)-1. It ignores the 'per year' convention and assumes compounding at every payment date. It is usually used to compare two long term opportunities.

Other Conventions and Uses

Exceptions:

- US and Canadian T-Bills (short term Government debt) have a different calculation for interest. Their interest is calculated as $(100–P)/P$ where 'P' is the price paid. Instead of normalizing it to a year, the interest is prorated by the number of days 't': (365/t)*100. The total calculation is $((100–P)/P)*((365/t)*100)$. This is equivalent to calculating the price by a process called discounting at a simple interest rate.
- Corporate Bonds are most frequently payable twice yearly. The amount of interest paid is the simple interest disclosed divided by two (multiplied by the face value of debt).

Flat Rate Loans and the Rule of .78s: Some consumer loans have been structured as flat rate loans, with the loan outstanding determined by allocating the total interest across the term of the loan by using the "Rule of 78s" or "Sum of digits" method. Seventy-eight is the sum of the numbers 1 through 12, inclusive. The practice enabled quick calculations of interest in the pre-computer days. In a loan with interest calculated per the Rule of 78s, the total interest over the life of the loan is calculated as either simple or compound interest and amounts to the same as either of the above methods. Payments remain constant over the life of the loan; however, payments are allocated to interest in progressively smaller amounts. In a one-year loan, in the first month, 12/78 of all interest owed over the life of the loan is due; in the second month, 11/78; progressing to the twelfth month where only 1/78 of all interest is due.

The practical effect of the Rule of 78s is to make early pay-offs of term loans more expensive. For a one year loan, approximately 3/4 of all interest due is collected by the sixth month, and pay-off of the principal then will cause the effective interest rate to be much higher than the APY used to calculate the payments.

In 1992, the United States outlawed the use of "Rule of 78s" interest in connection with mortgage refinancing and other consumer loans over five years in term. Certain other jurisdictions have outlawed application of the Rule of 78s in certain types of loans, particularly consumer loans.

Rule of 72: The "Rule of 72" is a "quick and dirty" method for finding out how fast money doubles for a given interest rate. For example, if you have an interest rate of 6%, it will take 72/6 or 12 years for your money to double, compounding at 6%. This is an approximation that starts to break down above 10%.

Market Interest Rates

There are markets for investments (which include the money market, bond market, as well as retail financial institutions like banks) set interest rates. Each specific debt takes into account the following factors in determining its interest rate:

Opportunity Cost

Opportunity cost encompasses any other use to which the money could be put, including lending to others, investing elsewhere, holding cash (for safety, for example), and simply spending the funds.

Inflation

Since the lender is deferring consumption, they will *wish*, as a bare minimum, to recover enough to pay the increased cost of goods due to inflation. Because future inflation is unknown, there are three ways this might be achieved:

- Charge *X*% interest 'plus inflation' Many governments issue 'real-return' or 'inflation indexed' bonds. The principal amount or the interest payments are continually increased by the rate of inflation.

- Decide on the 'expected' inflation rate. This still leaves the lender exposed to the risk of 'unexpected' inflation.
- Allow the interest rate to be periodically changed. While a 'fixed interest rate' remains the same throughout the life of the debt, 'variable' or 'floating' rates can be reset. There are derivative products that allow for hedging and swaps between the two.

However interest rates are set by the market, and it happens frequently that they are insufficient to compensate for inflation: for example at times of high inflation during e.g. the oil crisis; and currently (2011) when real yields on many inflation-linked government stocks are negative.

Default

There is always the risk the borrower will become bankrupt, abscond or otherwise default on the loan. The risk premium attempts to measure the integrity of the borrower, the risk of his enterprise succeeding and the security of any collateral pledged. For example, loans to developing countries have higher risk premiums than those to the US government due to the difference in creditworthiness. An operating line of credit to a business will have a higher rate than a mortgage loan.

The creditworthiness of businesses is measured by bond rating services and individual's credit scores by credit bureaus. The risks of an individual debt may have a large standard deviation of possibilities. The lender may want to cover his maximum risk, but lenders with portfolios of debt can lower the risk premium to cover just the most probable outcome.

Default Interest

Default interest is the interest that a borrower would pay if the borrower will not fulfill the loan covenants. The default interest is usually much higher than the original interest since it is reflecting the aggravation in the financial risk of the borrower. The default interest compensates the lender for the added risk.

Banks tend to add default interest to the loan agreements in order to separate between different scenarios.

Deferred Consumption

Charging interest equal only to inflation will leave the lender with the same purchasing power, but they would prefer their own consumption sooner rather than later. There will be an interest premium of the delay. They may not want to consume, but instead would invest in another product. The possible return they could realize in competing investments will determine what interest they charge.

Length of time

Shorter terms often have less risk of default and exposure to inflation because the near future is easier to predict. In these circumstances, short term interest rates are lower than longer term interest rates (an upward sloping yield curve).

Government Intervention

Interest rates are generally determined by the market, but government intervention – usually by a central bank - may strongly influence short-term interest rates, and is one of the main tools of monetary policy. The central bank offers to borrow (or lend) large quantities of money at a rate which they determine (sometimes this is money that they have created *ex nihilo*, i.e. printed) which has a major influence on supply and demand and hence on market interest rates.

Open Market Operations in the United States

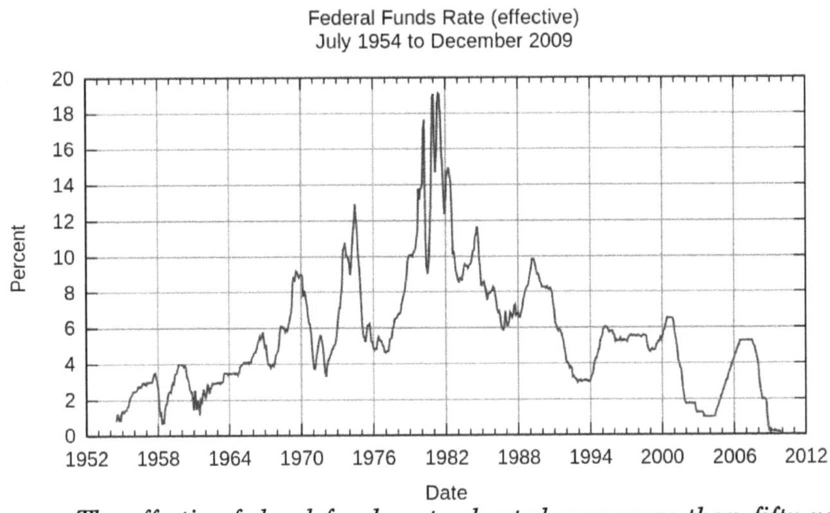

Figure: *The effective federal funds rate charted over more than fifty years.*

The Federal Reserve (Fed) implements monetary policy largely by targeting the federal funds rate. This is the rate that banks charge each other for overnight loans of federal funds. Federal funds are the reserves held by banks at the Fed.

Open market operations are one tool within monetary policy implemented by the Federal Reserve to steer short-term interest rates. Using the power to buy and sell treasury securities, the Open Market Desk at the Federal Reserve

Bank of New York can supply the market with dollars by purchasing T-notes, hence increasing the nation's money supply. By increasing the money supply or Aggregate Supply of Funding (ASF), interest rates will fall due to the excess of dollars banks will end up with in their reserves. Excess reserves may be lent in the Fed funds market to other banks, thus driving down rates.

Interest rates and credit risk

It is increasingly recognized that the business cycle, interest rates and credit risk are tightly interrelated. The Jarrow-Turnbull model was the first model of credit risk that explicitly had random interest rates at its core.

Lando (2004), Darrell Duffie and Singleton (2003), and van Deventer and Imai (2003) discuss interest rates when the issuer of the interest-bearing instrument can default.

Money and Inflation

Loans and bonds have some of the characteristics of money and are included in the broad money supply.

National governments (provided, of course, that the country has retained its own currency) can influence interest rates and thus the supply and demand for such loans, thus altering the total of loans and bonds issued. Generally speaking, a higher real interest rate reduces the broad money supply.

Through the quantity theory of money, increases in the money supply lead to inflation. This means that interest rates can affect inflation in the future.

Interest in Mathematics

It is thought that Jacob Bernoulli discovered the mathematical constant e by studying a question about compound interest. He realized that if an account that starts with \$1.00 and pays say 100% interest per year, at the end of the year, the value is \$2.00; but if the interest is computed and added twice in the year, the \$1 is multiplied by 1.5 twice, yielding $\$1.00 \times 1.5^2 = \$2.25$. Compounding quarterly yields $\$1.00 \times 1.25^4 = \$2.4414...$, and so on.

Bernoulli noticed that if the frequency of compounding is increased without limit, this sequence can be modelled as follows:

$$\lim_{n\to\infty}\left(1+\frac{1}{n}\right)^n = e,$$

where n is the number of times the interest is to be compounded in a year.

Formulae

The balance of a loan with regular monthly payments is augmented by the monthly interest charge and decreased by the payment so

$$B_{k+1} = (1+r)B_k - p,$$

where

i = loan rate/100 = annual rate in decimal form (e.g. 10% = 0.10 The loan rate is the rate used to compute payments and balances.)

r = period rate = $i/12$ for monthly payments (customary usage for convenience)

B_0 = initial balance (loan principal)

B_k = balance after k payments

k = balance index

p = period (monthly) payment

By repeated substitution one obtains expressions for B_k, which are linearly proportional to B_0 and p and use of the formula for the partial sum of a geometric series results in

$$B_k = (1+r)^k B_0 - \frac{(1+r)^k - 1}{r} p.$$

A solution of this expression for p in terms of B_0 and B_n reduces to

$$p = r\left[\frac{(1+r)^n B_0 - B_n}{(1+r)^n - 1}\right]$$

To find the payment if the loan is to be finished in n payments one sets $B_n = 0$.

The PMT function found in spreadsheet programmes can be used to calculate the monthly payment of a loan:

$$p = PMT(rate, num, PV, FV,) = PMT(r, n, -B_0, B_n,)$$

An interest-only payment on the current balance would be

$$p_I = rB.$$

The total interest, I_T, paid on the loan is

$$I_T = np - B_0 .$$

The formulas for a regular savings programme are similar but the payments are added to the balances instead of being subtracted and the formula for the payment is the negative of the one above. These formulas are only approximate since actual loan balances are affected by rounding.

To avoid an underpayment at the end of the loan, the payment must be rounded up to the next cent. The final payment would then be $(1+r)B_{n-1}$.

Consider a similar loan but with a new period equal to k periods of the problem above. If r_k and p_k are the new rate and payment, we now have

$$B_k = B_0' = (1+r_k)B_0 - p_k .$$

Comparing this with the expression for B_k above we note that

$$r_k = (1+r)^k - 1$$

and

$$p_k = \frac{p}{r} r_k .$$

The last equation allows us to define a constant that is the same for both problems,

$$B^* = \frac{p}{r} = \frac{p_k}{r_k}$$

and B_k can be written as

$$B_k = (1+r_k)B_0 - r_k B^* .$$

Solving for r_k we find a formula for r_k involving known quantities and B_k, the balance after k periods,

$$r_k = \frac{B_0 - B_k}{B^* - B_0}$$

Since B_0 could be any balance in the loan, the formula works for any two balances separate by k periods and can be used to compute a value for the annual interest rate.

B^* is a scale invariant since it does not change with changes in the length of the period.

Rearranging the equation for B^* one gets a transformation coefficient (scale factor),

$$\lambda_k = \frac{p_k}{p} = \frac{r_k}{r} = \frac{(1+r)^k - 1}{r} = k[1 + \frac{(k-1)r}{2} + \cdots]$$

and we see that r and p transform in the same manner,

$$r_k = \lambda_k r$$

$$p_k = \lambda_k p$$

The change in the balance transforms likewise,

$$\Delta B_k = B' - B = (\lambda_k rB - \lambda_k p) = \lambda_k \Delta B$$

which gives an insight into the meaning of some of the coefficients found in the formulas above. The annual rate, r_{12}, assumes only one payment per year and is not an "effective" rate for monthly payments. With monthly payments the monthly interest is paid out of each payment and so should not be compounded and an annual rate of $12 \cdot r$ would make more sense. If one just made interest-only payments the amount paid for the year would be $12 \cdot r \cdot B_0$.

Substituting $p_k = r_k B^*$ into the equation for the B_k we get,

$$B_k = B_0 - r_k(B^* - B_0)$$

Since $B_n = 0$ we can solve for B^*,

$$B^* = B_0\left(\frac{1}{r_n} + 1\right).$$

Substituting back into the formula for the B_k shows that they are a linear function of the r_k and therefore the λ_k,

$$B_k = B_0\left(1 - \frac{r_k}{r_n}\right) = B_0\left(1 - \frac{\lambda_k}{\lambda_n}\right)$$

This is the easiest way of estimating the balances if the λ_k are known. Substituting into the first formula for B_k above and solving for λ_{k+1} we get,

$$\lambda_{k+1} = 1 + (1+r)\lambda_k$$

λ_0 and λ_n can be found using the formula for λ_k above or computing the λ_k recursively from $\lambda_0 = 0$ to λ_n.

Since $p = rB^*$ the formula for the payment reduces to,

$$p = \left(r + \frac{1}{\lambda_n}\right)B_0$$

and the average interest rate over the period of the loan is

$$r_{loan} = \frac{I_T}{nB_0} = r + \frac{1}{\lambda_n} - \frac{1}{n},$$

which is less than r if n>1.

Compound Interest

Compound interest arises when interest is added to the principal, so that, from that moment on, the interest that has been added also earns interest. This addition of interest to the principal is called *compounding*. A bank account, for example, may have its interest compounded every year: in this case, an account with \$1000 initial principal and 20% interest per year would have a balance of \$1200 at the end of the first year, \$1440 at the end of the second year, and so on.

In order to define an interest rate fully, and enable one to compare it with other interest rates, the interest rate *and* the compounding frequency must be disclosed. Since most people prefer to think of rates as a yearly percentage, many governments require financial institutions to disclose the equivalent yearly compounded interest rate on deposits or advances. For instance, the yearly rate for a loan with 1% interest per month is approximately 12.68% per annum (1.01^{12}–1). This equivalent yearly rate may be referred to as *annual percentage rate* (APR), *annual equivalent rate* (AER), *annual percentage yield*, *effective interest rate*, *effective annual rate*, and by other terms. When a fee is charged up front to obtain a loan, APR usually counts that cost as well as the compound interest in converting to the equivalent rate. These government requirements assist consumers to compare the actual costs of borrowing more easily.

For any given interest rate and compounding frequency, an "equivalent" rate for any different compounding frequency exists. Compound interest may be contrasted with simple interest, where interest is not added to the principal (there is no compounding). Compound interest is standard in finance and economics, and simple interest is used infrequently (although certain financial products may contain elements of simple interest).

Terminology

The effect of compounding depends on the frequency with which interest is compounded and the periodic interest rate which is applied. Therefore, in order to define accurately the amount to be paid under a legal contract with interest, the frequency of compounding (yearly, half-yearly, quarterly, monthly, daily, etc.) *and* the interest rate must be specified. Different conventions may be used from country to country, but in finance and economics the following usages are common:

Periodic rate: the interest that is charged (and subsequently compounded) for each period, divided by the amount of the principal. The periodic rate is used primarily for calculations, and is rarely used for comparison. The nominal annual rate or nominal interest rate is defined as the periodic rate multiplied by the number of compounding periods per year. For example, a monthly rate of 1% is equivalent to an annual nominal interest of 12%.

Effective annual rate: this reflects the effective rate as if annual compounding were applied: in other words it is the total accumulated interest that would be payable up to the end of one year, divided by the principal.

Economists generally prefer to use effective annual rates to allow for comparability. In finance and commerce, the nominal annual rate may however be the one quoted instead. When quoted together with the compounding frequency, a loan with a given nominal annual rate is fully specified (the effect of interest for a given loan scenario can be precisely determined), but the nominal rate cannot be directly compared with loans that have a different compounding frequency.

Loans and finance may have other "non-interest" charges, and the terms above do not attempt to capture these differences. Other terms such as annual percentage rate and annual percentage yield may have specific legal definitions and may or may not be comparable, depending on the jurisdiction.

The use of the terms above (and other similar terms) may be inconsistent, and vary according to local custom or marketing demands, for simplicity or for other reasons.

Exceptions

- US and Canadian T-Bills (short term Government debt) have a different convention. Their interest is calculated as $(100-P)/Pbnm$, where P is the price paid. Instead of normalizing it to a year, the interest is prorated by the number of days t: $(365/t)\times 100$.
- The interest on corporate bonds and government bonds is usually payable twice yearly. The amount of interest paid (each six months) is the disclosed interest rate divided by two

(multiplied by the principal). The yearly compounded rate is higher than the disclosed rate.

- Canadian mortgage loans are generally semi-annual compounding with monthly (or more frequent) payments.
- U.S. mortgages use an amortizing loan, not compound interest. With these loans, an amortization schedule is used to determine how to apply payments towards principal and interest. Interest generated on these loans is not added to the principal, but rather is paid off monthly as the payments are applied.
- It is sometimes mathematically simpler, e.g. in the valuation of derivatives to use *continuous compounding*, which is the limit as the compounding period approaches zero. Continuous compounding in pricing these instruments is a natural consequence of *Itτ* calculus, where derivatives are valued at ever increasing frequency, until the limit is approached and the derivative is valued in continuous time.

Mathematics of Interest Rates

Simplified Calculation: Formulae are presented in greater detail at time value of money. In the formulae below, i is the effective interest rate per period. FV and PV represent the future and present value of a sum . n represents the number of periods.

These are the most basic formulas:

$$FV = PV(1+i)^n$$

The above calculates the future value (FV) of an investment's present value (PV) accruing at a fixed interest rate (i) for n periods.

$$PV = \frac{FV}{(1+i)^n}$$

The above calculates what present value (PV) would be needed to produce a certain future value (FV) if interest (i) accrues for n periods.

$$i = \left(\frac{FV}{PV}\right)^{\frac{1}{n}} - 1$$

The above calculates the compound interest rate achieved if an initial investment of PV returns a value of FV after n accrual periods.

$$n = \frac{\log(FV) - \log(PV)}{\log(1+i)}.$$

The above formula calculates the number of periods required to get *FV* given the *PV* and the interest rate (i). The log function can be in any base, e.g. natural log (ln), as long as consistent bases are used all throughout calculation.

Compound Interest

A formula for calculating annual compound interest is

$$A = P\left(1+\frac{r}{n}\right)^{nt}$$

Where,

- A = future value
- P = principal amount (initial investment)
- r = annual nominal interest rate
- n = number of times the interest is compounded per year
- t = number of years

Example usage: An amount of 1500.00 is deposited in a bank paying an annual interest rate of 4.3%, compounded quarterly. Find the balance after 6 years.

A. Using the formula above, with P = 1500, r = 0.043 (4.3%), n = 4, and t = 6:

$$A = 1500\left(1+\frac{0.043}{4}\right)^{4\times 6} = 1938.84$$

So, the balance after 6 years is approximately 1,938.84. Compound interest can be calculated by subtracting the principal from this balance or amount.

Periodic Compounding

The amount function for compound interest is a power law function in terms of time.

$$A(t) = A_0\left(1+\frac{r}{n}\right)^{\lfloor nt \rfloor}$$

- t= Total time in years
- n= Number of compounding periods per year (note that the total number of compounding periods is $n.t$)
- r= Nominal annual interest rate expressed as a decimal. e.g.: 6% = 0.06

- $\lfloor nt \rceil$ means that nt is rounded down to the nearest integer.

As n increases, the rate approaches an upper limit of e^{r-1}. This rate is called *continuous compounding.*

Since the principal $A(0)$ is simply a coefficient, it is often dropped for simplicity, and the resulting accumulation function is used in interest theory instead. Accumulation functions for simple and compound interest are listed below:

$$a(t) = 1 + tr$$

$$a(t) = \left(1 + \frac{r}{n}\right)^{nt}$$

Note: $A(t)$ is the amount function and $a(t)$ is the accumulation function.

Continuous Compounding

Continuous compounding can be thought of as making the compounding period infinitesimally small; therefore achieved by taking the limit of n to infinity.

One should consult definitions of the exponential function for the mathematical proof of this limit.

$$A(t) = A_0 e^{rt}$$

or

$$A = Pe^{rt}$$

Force of Interest

In mathematics, the accumulation functions are often expressed in terms of e, the base of the natural logarithm. This facilitates the use of calculus methods in manipulation of interest formulae.

For any continuously differentiable accumulation function*a(t)* the force of interest, or more generally the logarithmic or continuously compounded return is a function of time defined as follows:

$$\delta_t = \frac{a'(t)}{a(t)}$$

which is the rate of change with time of the natural logarithm of the accumulation function.

Conversely:

$$a(n) = e^{\int_0^n \delta_t \, dt}, \quad (\text{since } a(0) = 1)$$

When the above formula is written in differential equation format, the force of interest is simply the coefficient of amount of change: $da(t) = \delta_t a(t)dt$

For compound interest with a constant annual interest rate r, the force of interest is a constant, and the accumulation function of compounding interest in terms of force of interest is a simple power of e: $\delta = \ln(1+r)$ or $a(t) = e^{t\delta}$

The force of interest is less than the annual effective interest rate, but more than the annual effective discount rate. It is the reciprocal of the e-folding time.

A way of modelling the force of inflation is with Stoodley's formula: $\delta_t = p + \frac{s}{1+rse^{st}}$ where p, r and s are estimated.

Compounding Basis

To convert an interest rate from one compounding basis to another compounding basis, the following formula applies:

$$r_2 = \left[\left(1+\frac{r_1}{n_1}\right)^{\frac{n_1}{n_2}} - 1\right] n_2$$

where r_1 is the stated interest rate with compounding frequency n_1 and r_2 is the stated interest rate with compounding frequency n_2.

When interest is continuously compounded:

$$R = n\ln(1 + r/n)$$

where R is the interest rate on a continuous compounding basis and r is the stated interest rate with a compounding frequency n.

Monthly Mortgage Payments

The interest on mortgages is often compounded monthly. The formula for payments is found from the following argument.

Notation

- I = Note percentage rate
- i = Monthly percentage rate = $I/12$ (so that the APR = $(1+i)^12-1$)
- T = Term in years
- $Y = I \cdot T$

- $X = \frac{1}{2} I \cdot T = \frac{1}{2} Y$
- $n = 12 \cdot T$ = term in months
- L = Principal or amount of loan
- P = monthly payment.

History

Compound interest was once regarded as the worst kind of usury, and was severely condemned by Roman law, as well as the common laws of many other countries.

In one passage, the Bible addresses the charging of interest in the following manner:

Take no usury or interest from him; but fear your God, that your brother may live with you. You shall not lend him your money for usury, nor lend him your food at a profit.

The Qur'an explicitly mentions compound interest as a great sin. Usury (oppressive interest), known in Arabic as "riba", is considered wrong:

O ye who believe! Devour not usury, doubling and quadrupling (the sum lent). Observe your duty to Allah, that ye may be successful.

— Quran 3:130

Richard Witt's book *Arithmeticall Questions*, published in 1613, was a landmark in the history of compound interest. It was wholly devoted to the subject (previously called anatocism), whereas previous writers had usually treated compound interest briefly in just one chapter in a mathematical textbook. Witt's book gave tables based on 10% (the then maximum rate of interest allowable on loans) and on other rates for different purposes, such as the valuation of property leases. Witt was a London mathematical practitioner and his book is notable for its clarity of expression, depth of insight and accuracy of calculation, with 124 worked examples.

Chapter

6

The Sum of Two Quadratics

Scalar Form

The following auxiliary formula is useful for simplifying the posterior update equations, which otherwise become fairly tedious.

$$a(y-x)^2+b(x-z)^2=(a+b)\left(x-\frac{ay+bz}{a+b}\right)^2+\frac{ab}{a+b}(y-z)^2$$

This equation rewrites the sum of two quadratics in x by expanding the squares, grouping the terms in x, and completing the square.

Note the following about the complex constant factors attached to some of the terms:

1. The factor $\frac{ay+bz}{a+b}$ has the form of a weighted average of y and z.
2. $\frac{ab}{a+b}=\frac{1}{\frac{1}{a}+\frac{1}{b}}=(a^{-1}+b^{-1})^{-1}.$

   This shows that this factor can be thought of as resulting from a situation where the reciprocals of quantities a and b add directly, so to combine a and b themselves, it's necessary to reciprocate, add, and reciprocate the result again to get back into the original units.

This is exactly the sort of operation performed by the harmonic mean, so it is not surprising that $\frac{ab}{a+b}$ is one-half the harmonic mean of a and b.

Vector form

A similar formula can be written for the sum of two vector quadratics: If x, y, z are vectors of length k, and A and B are symmetric, invertible matrices of size $k \times k$, then

$$\begin{aligned}&(y-x)'A(y-x)+(x-z)'B(x-z)\\&=(x-c)'(A+B)(x-c)+(y-z)'(A^{-1}+B^{-1})^{-1}(y-z)\end{aligned}$$

where

$$c=(A+B)^{-1}(Ay+Bz)$$

Note that the form x2 Ax is called a quadratic form and is a scalar:

$$x'Ax=\sum_{i,j}a_{ij}x_i x_j$$

In other words, it sums up all possible combinations of products of pairs of elements from x, with a separate coefficient for each. In addition, since $x_i x_j = x_j x_i$, only the sum $a_{ij}+a_{ji}$ matters for any off-diagonal elements of A, and there is no loss of generality in assuming that A is symmetric.

Furthermore, if A is symmetric, then the form $x'Ay = y'Ax$.

The Sum of Differences from the Mean

Another useful formula is as follows:

$$\sum_{i=1}^{n}(x_i-\mu)^2=\sum_{i=1}^{n}(x_i-\overline{x})^2+n(\overline{x}-\mu)^2$$

where $\overline{x}=\frac{1}{n}\sum_{i=1}^{n}x_i$.

With Known Variance

For a set of i.i.d. normally distributed data points X of size n where each individual point x follows $x \sim \mathcal{N}(\mu,\sigma^2)$ with known variance σ^2, the conjugate prior distribution is also normally distributed. This can be shown more easily by rewriting the variance as the precision, i.e. using $\tau = 1/\sigma^2$.

Then if $x \sim \mathcal{N}(\mu,\tau)$ and $\mu \sim \mathcal{N}(\mu_0,\tau_0)$, we proceed as follows.

First, the likelihood function is (using the formula above for the sum of differences from the mean):

$$\begin{aligned} p(X \mid \mu,\tau) &= \prod_{i=1}^{n} \sqrt{\frac{\tau}{2\pi}} \exp\left(-\frac{1}{2}\tau(x_i-\mu)^2\right) \\ &= \left(\frac{\tau}{2\pi}\right)^{\frac{n}{2}} \exp\left(-\frac{1}{2}\tau\sum_{i=1}^{n}(x_i-\mu)^2\right) \\ &= \left(\frac{\tau}{2\pi}\right)^{\frac{n}{2}} \exp\left[-\frac{1}{2}\tau\left(\sum_{i=1}^{n}(x_i-\bar{x})^2+n(\bar{x}-\mu)^2\right)\right]. \end{aligned}$$

Then, we proceed as follows:

$$\begin{aligned} p(\mu \mid X) &\propto p(X \mid \mu)p(\mu) \\ &= \left(\frac{\tau}{2\pi}\right)^{\frac{n}{2}} \exp\left[-\frac{1}{2}\tau\left(\sum_{i=1}^{n}(x_i-\bar{x})^2+n(\bar{x}-\mu)^2\right)\right]\sqrt{\frac{\tau_0}{2\pi}}\exp\left(-\frac{1}{2}\tau_0(\mu-\mu_0)^2\right) \\ &\propto \exp\left(-\frac{1}{2}\left(\tau\left(\sum_{i=1}^{n}(x_i-\bar{x})^2+n(\bar{x}-\mu)^2\right)+\tau_0(\mu-\mu_0)^2\right)\right) \\ &\propto \exp\left(-\frac{1}{2}\left(n\tau(\bar{x}-\mu)^2+\tau_0(\mu-\mu_0)^2\right)\right) \\ &= \exp\left(-\frac{1}{2}(n\tau+\tau_0)\left(\mu-\frac{n\tau\bar{x}+\tau_0\mu_0}{n\tau+\tau_0}\right)^2+\frac{n\tau\tau_0}{n\tau+\tau_0}(\bar{x}-\mu_0)^2\right) \\ &\propto \exp\left(-\frac{1}{2}(n\tau+\tau_0)\left(\mu-\frac{n\tau\bar{x}+\tau_0\mu_0}{n\tau+\tau_0}\right)^2\right) \end{aligned}$$

In the above derivation, we used the formula above for the sum of two quadratics and eliminated all constant factors not involving μ.

The result is the kernel of a normal distribution, with mean

$$\frac{n\tau\bar{x}+\tau_0\mu_0}{n\tau+\tau_0}$$

and precision $n\tau+\tau_0$, i.e.

$$p(\mu \mid X) \sim \mathcal{N}\left(\frac{n\tau\bar{x}+\tau_0\mu_0}{n\tau+\tau_0}, n\tau+\tau_0\right)$$

This can be written as a set of Bayesian update equations for the posterior parameters in terms of the prior parameters:

$$\tau_{0'} = \tau_0 + n\tau$$

$$\mu_{0'} = \frac{n\tau\bar{x} + \tau_0\mu_0}{n\tau + \tau_0}$$

$$\bar{x} = \frac{1}{n}\sum_{i=1}^{n} x_i$$

That is, to combine n data points with total precision of $n\tau$ (or equivalently, total variance of n/σ^2) and mean of values $\bar{x}$, derive a new total precision simply by adding the total precision of the data to the prior total precision, and form a new mean through a *precision-weighted average*, i.e. a weighted average of the data mean and the prior mean, each weighted by the associated total precision. This makes logical sense if the precision is thought of as indicating the certainty of the observations:

In the distribution of the posterior mean, each of the input components is weighted by its certainty, and the certainty of this distribution is the sum of the individual certainties. (For the intuition of this, compare the expression "the whole is (or is not) greater than the sum of its parts". In addition, consider that the knowledge of the posterior comes from a combination of the knowledge of the prior and likelihood, so it makes sense that we are more certain of it than of either of its components.)

The above formula reveals why it is more convenient to do Bayesian analysis of conjugate priors for the normal distribution in terms of the precision. The posterior precision is simply the sum of the prior and likelihood precisions, and the posterior mean is computed through a precision-weighted average, as described above. The same formulas can be written in terms of variance by reciprocating all the precisions, yielding the more ugly formulas

$$\sigma_0^{2'} = \frac{1}{\frac{n}{\sigma^2} + \frac{1}{\sigma_0^2}}$$

$$\mu_{0'} = \frac{\frac{n\bar{x}}{\sigma^2} + \frac{\mu_0}{\sigma_0^2}}{\frac{n}{\sigma^2} + \frac{1}{\sigma_0^2}}$$

$$\bar{x} = \frac{1}{n}\sum_{i=1}^{n} x_i$$

With Known Mean

For a set of i.i.d. normally distributed data points X of size n where each individual point x follows $x \sim \mathcal{N}(\mu,\sigma^2)$ with known mean μ, the conjugate prior of the variance has an inverse gamma distribution or a scaled inverse chi-squared distribution. The two are equivalent except for having different parameterizations. The use of the inverse gamma is more common, but the scaled inverse chi-squared is more convenient, so we use it in the following derivation. The prior for σ^2 is as follows:

$$p(\sigma^2 \mid \nu_0,\sigma_0^2) = \frac{(\sigma_0^2 \frac{\nu_0}{2})^{\frac{\nu_0}{2}}}{\Gamma\left(\frac{\nu_0}{2}\right)} \frac{\exp\left[\frac{-\nu_0\sigma_0^2}{2\sigma^2}\right]}{(\sigma^2)^{1+\frac{\nu_0}{2}}} \propto \frac{\exp\left[\frac{-\nu_0\sigma_0^2}{2\sigma^2}\right]}{(\sigma^2)^{1+\frac{\nu_0}{2}}}$$

The likelihood function from above, written in terms of the variance, is:

$$\begin{aligned} p(\mathrm{X} \mid \mu,\sigma^2) &= \left(\frac{1}{2\pi\sigma^2}\right)^{\frac{n}{2}} \exp\left[-\frac{1}{2\sigma^2}\sum_{i=1}^{n}(x_i-\mu)^2\right] \\ &= \left(\frac{1}{2\pi\sigma^2}\right)^{\frac{n}{2}} \exp\left[-\frac{S}{2\sigma^2}\right] \end{aligned}$$

where

$$S = \sum_{i=1}^{n}(x_i-\mu)^2.$$

Then:

$$\begin{aligned} p(\sigma^2 \mid \mathrm{X}) &\propto p(\mathrm{X} \mid \sigma^2)p(\sigma^2) \\ &= \left(\frac{1}{2\pi\sigma^2}\right)^{\frac{n}{2}} \exp\left[-\frac{S}{2\sigma^2}\right] \frac{(\sigma_0^2 \frac{\nu_0}{2})^{\frac{\nu_0}{2}}}{\Gamma\left(\frac{\nu_0}{2}\right)} \frac{\exp\left[\frac{-\nu_0\sigma_0^2}{2\sigma^2}\right]}{(\sigma^2)^{1+\frac{\nu_0}{2}}} \\ &\propto \left(\frac{1}{\sigma^2}\right)^{\frac{n}{2}} \frac{1}{(\sigma^2)^{1+\frac{\nu_0}{2}}} \exp\left[-\frac{S}{2\sigma^2} + \frac{-\nu_0\sigma_0^2}{2\sigma^2}\right] \\ &= \frac{1}{(\sigma^2)^{1+\frac{\nu_0+n}{2}}} \exp\left[-\frac{\nu_0\sigma_0^2+S}{2\sigma^2}\right] \end{aligned}$$

This is also a scaled inverse chi-squared distribution, where

$$\nu_{0'} = \nu_0 + n$$

$$\nu_{0'}\sigma_0^{2'} = \nu_0\sigma_0^2 + \sum_{i=1}^{n}(x_i - \mu)^2$$

or equivalently

$$\nu_{0'} = \nu_0 + n$$

$$\sigma_0^{2'} = \frac{\nu_0\sigma_0^2 + \sum_{i=1}^{n}(x_i - \mu)^2}{\nu_0 + n}$$

Reparameterizing in terms of an inverse gamma distribution, the result is:

$$\alpha' = \alpha + \frac{n}{2}$$

$$\beta' = \beta + \frac{\sum_{i=1}^{n}(x_i - \mu)^2}{2}$$

With Unknown Mean and Variance

For a set of i.i.d. normally distributed data points X of size n where each individual point x follows $x \sim \mathcal{N}(\mu, \sigma^2)$ with unknown mean μ and variance σ^2, a combined (multivariate) conjugate prior is placed over the mean and variance, consisting of a normal-inverse-gamma distribution.

Logically, this originates as follows:

1. From the analysis of the case with unknown mean but known variance, we see that the update equations involve sufficient statistics computed from the data consisting of the mean of the data points and the total variance of the data points, computed in turn from the known variance divided by the number of data points.
2. From the analysis of the case with unknown variance but known mean, we see that the update equations involve sufficient statistics over the data consisting of the number of data points and sum of squared deviations.
3. Keep in mind that the posterior update values serve as the prior distribution when further data is handled. Thus, we should logically think of our priors in terms of the sufficient

statistics just described, with the same semantics kept in mind as much as possible.

4. To handle the case where both mean and variance are unknown, we could place independent priors over the mean and variance, with fixed estimates of the average mean, total variance, number of data points used to compute the variance prior, and sum of squared deviations.

   Note however that in reality, the total variance of the mean depends on the unknown variance, and the sum of squared deviations that goes into the variance prior (appears to) depend on the unknown mean.

   In practice, the latter dependence is relatively unimportant: Shifting the actual mean shifts the generated points by an equal amount, and on average the squared deviations will remain the same. This is not the case, however, with the total variance of the mean: As the unknown variance increases, the total variance of the mean will increase proportionately, and we would like to capture this dependence.

5. This suggests that we create a *conditional prior* of the mean on the unknown variance, with a hyperparameter specifying the mean of the pseudo-observations associated with the prior, and another parameter specifying the number of pseudo-observations. This number serves as a scaling parameter on the variance, making it possible to control the overall variance of the mean relative to the actual variance parameter.

   The prior for the variance also has two hyperparameters, one specifying the sum of squared deviations of the pseudo-observations associated with the prior, and another specifying once again the number of pseudo-observations. Note that each of the priors has a hyperparameter specifying the number of pseudo-observations, and in each case this controls the relative variance of that prior. These are given as two separate hyperparameters so that the variance (aka the confidence) of the two priors can be controlled separately.

6. This leads immediately to the normal-inverse-gamma distribution, which is defined as the product of the two distributions just defined, with conjugate priors used (an inverse gamma distribution over the variance, and a normal distribution over the mean, *conditional* on the variance) and with the same four parameters just defined.

The priors are normally defined as follows:

$$p(\mu \mid \sigma^2; \mu_0, n_0) \sim \mathcal{N}(\mu_0, \sigma_0^2 / n_0)$$

$$p(\sigma^2; \nu_0, \sigma_0^2) \sim I\chi^2(\nu_0, \sigma_0^2) = IG(\nu_0 / 2, \nu_0\sigma_0^2 / 2)$$

The update equations can be derived, and look as follows:

$$\bar{x} = \frac{1}{n}\sum_{i=1}^{n} x_i$$

$$\mu_{0'} = \frac{n_0\mu_0 + n\bar{x}}{n_0 + n}$$

$$n_{0'} = n_0 + n$$

$$\nu_{0'} = \nu_0 + n$$

$$\nu_{0'}\sigma_0^{2'} = \nu_0\sigma_0^2 + \sum_{i=1}^{n}(x_i - \bar{x})^2 + \frac{n_0 n}{n_0 + n}(\mu_0 - \bar{x})^2$$

The respective numbers of pseudo-observations just add the number of actual observations to them. The new mean hyperparameter is once again a weighted average, this time weighted by the relative numbers of observations. Finally, the update for $\nu_{0'}\sigma_0^{2'}$ is similar to the case with known mean, but in this case the sum of squared deviations is taken with respect to the observed data mean rather than the true mean, and as a result a new "interaction term" needs to be added to take care of the additional error source stemming from the deviation between prior and data mean.

Proof is as follows.

Occurrence

The occurrence of normal distribution in practical problems can be loosely classified into three categories:

1. Exactly normal distributions;
2. Approximately normal laws, for example when such approximation is justified by the central limit theorem; and
3. Distributions modelled as normal – the normal distribution being the distribution with maximum entropy for a given mean and variance.

Exact Normality

Certain quantities in physics are distributed normally, as was first demonstrated by James Clerk Maxwell. Examples of such quantities are:

- Velocities of the molecules in the ideal gas. More generally, velocities of the particles in any system in thermodynamic equilibrium will have normal distribution, due to the maximum entropy principle.
- Probability density function of a ground state in a quantum harmonic oscillator.
- The position of a particle which experiences diffusion. If initially the particle is located at a specific point (that is its probability distribution is the dirac delta function), then after time *t* its location is described by a normal distribution with variance *t*, which satisfies the diffusion equation ∂∂*t* *f*(*x*,*t*) = 12 $\partial^2\partial x^2$ *f*(*x*,*t*). If the initial location is given by a certain density function *g*(*x*), then the density at time *t* is the convolution of *g* and the normal PDF.

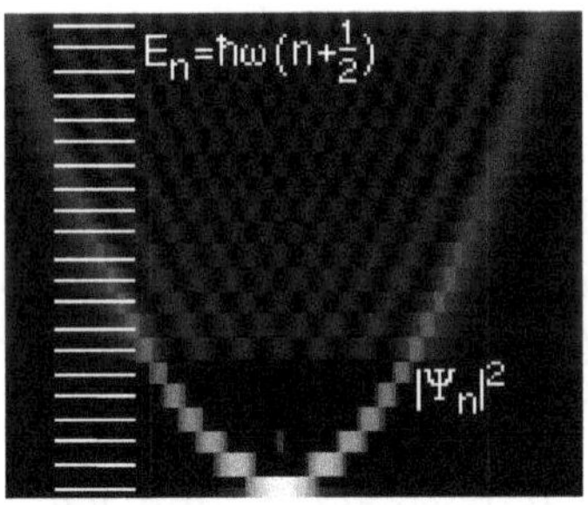

Figure: *The ground state of a quantum harmonic oscillator has the Gaussian distribution.*

Approximate Normality

Approximately normal distributions occur in many situations, as explained by the central limit theorem. When the outcome is produced by a large number of small effects acting *additively and independently*, its distribution will be close to normal. The normal approximation will not be valid if the effects act multiplicatively (instead of additively), or if there is a single external influence which has a considerably larger magnitude than the rest of the effects.

- In counting problems, where the central limit theorem includes a discrete-to-continuum approximation and where infinitely divisible and decomposable distributions are involved, such as
  - Binomial random variables, associated with binary response variables;
  - Poisson random variables, associated with rare events;
- Thermal light has a Bose–Einstein distribution on very short

time scales, and a normal distribution on longer timescales due to the central limit theorem.

Assumed Normality

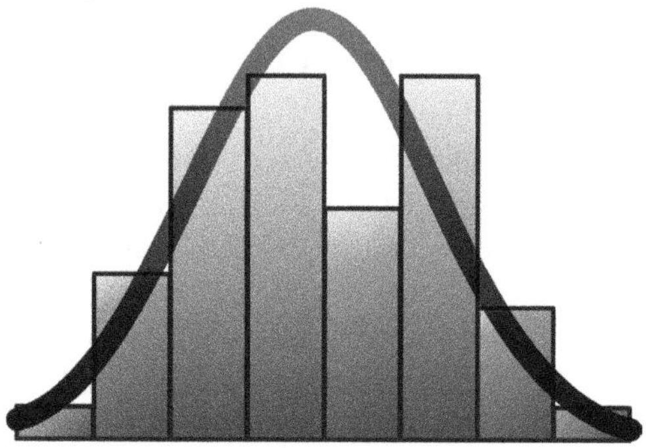

Figure: *Histogram of sepal widths for* Iris versicolour *from Fisher's Iris flower data set, with superimposed best-fitting normal distribution.*

I can only recognize the occurrence of the normal curve – the Laplacian curve of errors – as a very abnormal phenomenon. It is roughly approximated to in certain distributions; for this reason, and on account for its beautiful simplicity, we may, perhaps, use it as a first approximation, particularly in theoretical investigations.

—Pearson (1901)

There are statistical methods to empirically test that assumption.

- In biology, the *logarithm* of various variables tend to have a normal distribution, that is, they tend to have a log-normal distribution (after separation on male/female subpopulations), with examples including:
  - Measures of size of living tissue (length, height, skin area, weight);
  - The *length* of *inert* appendages (hair, claws, nails, teeth) of biological specimens, *in the direction of growth*; presumably the thickness of tree bark also falls under this category;
  - Certain physiological measurements, such as blood pressure of adult humans.
- In finance, in particular the Black–Scholes model, changes in the *logarithm* of exchange rates, price indices, and stock market indices are assumed normal (these variables behave like compound interest, not like simple interest, and so are multiplicative). Some mathematicians such as Benoît

Mandelbrot have argued that log-Levy distributions which possesses heavy tails would be a more appropriate model, in particular for the analysis for stock market crashes.

- Measurement errors in physical experiments are often modelled by a normal distribution. This use of a normal distribution does not imply that one is assuming the measurement errors are normally distributed, rather using the normal distribution produces the most conservative predictions possible given only knowledge about the mean and variance of the errors.

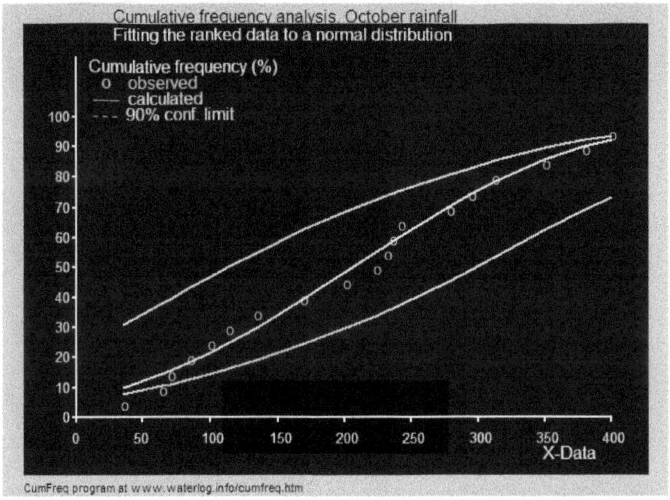

***Figure:** Fitted cumulative normal distribution to October rainfalls*

- In standardized testing, results can be made to have a normal distribution. This is done by either selecting the number and difficulty of questions (as in the IQ test), or by transforming the raw test scores into "output" scores by fitting them to the normal distribution. For example, the SAT's traditional range of 200–800 is based on a normal distribution with a mean of 500 and a standard deviation of 100.
- Many scores are derived from the normal distribution, including percentile ranks ("percentiles" or "quantiles"), normal curve equivalents, stanines, z-scores, and T-scores. Additionally, a number of behavioural statistical procedures are based on the assumption that scores are normally distributed; for example, t-tests and ANOVAs. Bell curve grading assigns relative grades based on a normal distribution of scores.
- In hydrology the distribution of long duration river discharge or rainfall, e.g. monthly and yearly totals, is often thought to

be practically normal according to the central limit theorem. The blue picture illustrates an example of fitting the normal distribution to ranked October rainfalls showing the 90% confidence belt based on the binomial distribution. The rainfall data are represented by plotting positions as part of the cumulative frequency analysis.

Generating Values from Normal Distribution

Figure: *The bean machine, a device invented by Francis Galton, can be called the first generator of normal random variables. This machine consists of a vertical board with interleaved rows of pins. Small balls are dropped from the top and then bounce randomly left or right as they hit the pins. The balls are collected into bins at the bottom and settle down into a pattern resembling the Gaussian curve.*

In computer simulations, especially in applications of the Monte-Carlo method, it is often desirable to generate values that are normally distributed. The algorithms listed below all generate the standard normal deviates, since a $N(\mu, \sigma 2)$ can be generated as $X = \mu + \sigma Z$, where Z is standard normal. All these algorithms rely on the availability of a random number generator U capable of producing uniform random variates.

- The most straight forward method is based on the probability integral transform property: if U is distributed uniformly on (0,1), then $\Phi^{-1}(U)$ will have the standard normal distribution. The drawback of this method is that it relies on calculation of the probit function Φ^{-1}, which cannot be done analytically.

Some approximate methods are described in Hart (1968) and in the *erf* article. Wichura gives a fast algorithm for computing this function to 16 decimal places, which is used by *R* to compute random variates of the normal distribution.

- An easy to programme approximate approach, that relies on the central limit theorem, is as follows: generate 12 uniform $U(0,1)$ deviates, add them all up, and subtract 6 – the resulting random variable will have approximately standard normal distribution. In truth, the distribution will be Irwin–Hall, which is a 12-section eleventh-order polynomial approximation to the normal distribution. This random deviate will have a limited range of (–6, 6).
- The Box–Muller method uses two independent random numbers U and V distributed uniformly on (0,1). Then the two random variables X and Y

$$X = \sqrt{-2\ln U}\cos(2\pi V),$$
$$Y = \sqrt{-2\ln U}\sin(2\pi V).$$

will both have the standard normal distribution, and will be independent. This formulation arises because for a bivariate normal random vector (XY) the squared norm $X^2 + Y^2$ will have the chi-squared distribution with two degrees of freedom, which is an easily generated exponential random variable corresponding to the quantity $-2\ln(U)$ in these equations; and the angle is distributed uniformly around the circle, chosen by the random variable V.

- Marsaglia polar method is a modification of the Box–Muller method algorithm, which does not require computation of functions sin() and cos(). In this method U and V are drawn from the uniform (–1,1) distribution, and then $S = U^2 + V^2$ is computed. If S is greater or equal to one then the method starts over, otherwise two quantities

$$X = U\sqrt{\frac{-2\ln S}{S}}, \qquad Y = V\sqrt{\frac{-2\ln S}{S}}$$

are returned. Again, X and Y will be independent and standard normally distributed.

- The Ratio method is a rejection method. The algorithm proceeds as follows:
  - Generate two independent uniform deviates U and V;

— Compute $X = \sqrt{8/e}\,(V - 0.5)/U$;
— Optional: if $X^2 \leq 5 - 4e^{1/4}U$ then accept X and terminate algorithm;
— Optional: if $X^2 \geq 4e^{-1.35}/U + 1.4$ then reject X and start over from step 1;
— If $X^2 \leq -4 \ln U$ then accept X, otherwise start over the algorithm.

- The ziggurat algorithm is faster than the Box–Muller transform and still exact. In about 97% of all cases it uses only two random numbers, one random integer and one random uniform, one multiplication and an if-test. Only in 3% of the cases where the combination of those two falls outside the "core of the ziggurat" a kind of rejection sampling using logarithms, exponentials and more uniform random numbers has to be employed.
- There is also some investigation into the connection between the fast Hadamard transform and the normal distribution, since the transform employs just addition and subtraction and by the central limit theorem random numbers from almost any distribution will be transformed into the normal distribution. In this regard a series of Hadamard transforms can be combined with random permutations to turn arbitrary data sets into a normally distributed data.

Numerical Approximations for the Normal CDF

The standard normal CDF is widely used in scientific and statistical computing. The values $\Phi(x)$ may be approximated very accurately by a variety of methods, such as numerical integration, Taylor series, asymptotic series and continued fractions. Different approximations are used depending on the desired level of accuracy.

- Zelen & Severo (1964) give the approximation for $\Phi(x)$ for $x > 0$ with the absolute error $|\varepsilon(x)| < 7.5.10^{-8}$ (algorithm 26.2.17):

$$\Phi(x) = 1 - \phi(x)\left(b_1 t + b_2 t^2 + b_3 t^3 + b_4 t^4 + b_5 t^5\right) + \varepsilon(x), \qquad t = \frac{1}{1 + b_0 x},$$

  where $\Phi(x)$ is the standard normal PDF, and $b_0 = 0.2316419$, $b_1 = 0.319381530$, $b_2 = -0.356563782$, $b_3 = 1.781477937$, $b_4 = -1.821255978$, $b_5 = 1.330274429$.
- Hart (1968) lists almost a hundred of rational function approximations for the erfc() function. His algorithms vary in

the degree of complexity and the resulting precision, with maximum absolute precision of 24 digits. An algorithm by West (2009) combines Hart's algorithm 5666 with a continued fraction approximation in the tail to provide a fast computation algorithm with a 16-digit precision.

- Cody (1969) after recalling Hart68 solution is not suited for *erf*, gives a solution for both *erf* and *erfc*, with maximal relative error bound, via Rational Chebyshev Approximation.
- Marsaglia (2004) suggested a simple algorithm based on the Taylor series expansion

$$\Phi(x) = \frac{1}{2} + \phi(x)\left(x + \frac{x^3}{3} + \frac{x^5}{3 \cdot 5} + \frac{x^7}{3 \cdot 5 \cdot 7} + \frac{x^9}{3 \cdot 5 \cdot 7 \cdot 9} + \cdots \right)$$

  for calculating $\Phi(x)$ with arbitrary precision. The drawback of this algorithm is comparatively slow calculation time (for example it takes over 300 iterations to calculate the function with 16 digits of precision when $x = 10$).
- The GNU Scientific Library calculates values of the standard normal CDF using Hart's algorithms and approximations with Chebyshev polynomials.

History

Development: Some authors attribute the credit for the discovery of the normal distribution to de Moivre, who in 1738 published in the second edition of his "*The Doctrine of Chances*" the study of the coefficients in the binomial expansion of $(a + b)^n$. De Moivre proved that the middle term in this expansion has the approximate magnitude of $2/\sqrt{2\pi n}$, and that "If m or ½n be a Quantity infinitely great, then the Logarithm of the Ratio, which a Term distant from the middle by the Interval ℓ, has to the middle Term, is $-\frac{2\ell\ell}{n}$." Although this theorem can be interpreted as the first obscure expression for the normal probability law, Stigler points out that de Moivre himself did not interpret his results as anything more than the approximate rule for the binomial coefficients, and in particular de Moivre lacked the concept of the probability density function.

In 1809 Gauss published his monograph "*The oriamotus corporum coelestium in sectionibus conicis solem ambientium*" where among other things he introduces several important statistical concepts, such as the method of least squares, the method of maximum likelihood, and the *normal distribution*. Gauss used $M, M2$, $M2$ 2 , ... to denote the measurements of some unknown quantity V, and sought the "most probable" estimator: the one which maximizes the probability $\varphi(M-V)\cdot\varphi(M'-V)\cdot\varphi(M''-V)\cdot\ldots$ of obtaining the observed experimental results.

In his notation $\varphi\Delta$ is the probability law of the measurement errors of magnitude Δ. Not knowing what the function φ is, Gauss requires that his method should reduce to the well-known answer: the arithmetic mean of the measured values. Starting from these principles, Gauss demonstrates that the only law which rationalizes the choice of arithmetic mean as an estimator of the location parameter, is the normal law of errors:

$$\varphi\Delta = \frac{h}{\sqrt{\pi}} e^{-hh\Delta\Delta},$$

where$h$ is "the measure of the precision of the observations". Using this normal law as a generic model for errors in the experiments, Gauss formulates what is now known as the non-linear weighted least squares (NWLS) method.

Although Gauss was the first to suggest the normal distribution law, Laplace made significant contributions. It was Laplace who first posed the problem of aggregating several observations in 1774, although his own solution led to the Laplacian distribution.

It was Laplace who first calculated the value of the integral $\int e^{-t^2} dt = \sqrt{\pi}$ in 1782, providing the normalization constant for the normal distribution. Finally, it was Laplace who in 1810 proved and presented to the Academy the fundamental central limit theorem, which emphasized the theoretical importance of the normal distribution.

It is of interest to note that in 1809 an American mathematician Adrain published two derivations of the normal probability law, simultaneously and independently from Gauss. His works remained largely unnoticed by the scientific community, until in 1871 they were "rediscovered" by Abbe.

In the middle of the 19th century Maxwell demonstrated that the normal distribution is not just a convenient mathematical tool, but may also occur in natural phenomena: "The number of particles whose velocity, resolved in a certain direction, lies between x and $x + dx$ is

$$N \frac{1}{\alpha\sqrt{\pi}} e^{-\frac{x^2}{\alpha^2}} dx$$

Naming

Since its introduction, the normal distribution has been known by many different names: the law of error, the law of facility of errors, Laplace's second law, Gaussian law, etc. Gauss himself apparently coined the term with reference to the "normal equations" involved in its applications, with normal having its technical meaning of orthogonal rather than "usual".

However, by the end of the 19th century some authors had started using the name *normal distribution*, where the word "normal" was used as an adjective – the term now being seen as a reflection of the fact that this distribution was seen as typical, common – and thus "normal". Peirce (one of those authors) once defined "normal" thus: "...the 'normal' is not the average (or any other kind of mean) of what actually occurs, but of what *would*, in the long run, occur under certain circumstances." Around the turn of the 20th century Pearson popularized the term *normal* as a designation for this distribution.

> Many years ago I called the Laplace–Gaussian curve the *normal* curve, which name, while it avoids an international question of priority, has the disadvantage of leading people to believe that all other distributions of frequency are in one sense or another 'abnormal'.
>
> —Pearson (1920)

Also, it was Pearson who first wrote the distribution in terms of the standard deviation σ as in modern notation. Soon after this, in year 1915, Fisher added the location parameter to the formula for normal distribution, expressing it in the way it is written nowadays:

$$df = \frac{1}{\sigma\sqrt{2\pi}} e^{-\frac{(x-m)^2}{2\sigma^2}} dx$$

The term "standard normal" which denotes the normal distribution with zero mean and unit variance came into general use around 1950s, appearing in the popular textbooks by P.G. Hoel (1947) "*Introduction to mathematical statistics*" and A.M. Mood (1950) "*Introduction to the theory of statistics*".

When the name is used, the "Gaussian distribution" was named after Carl Friedrich Gauss, who introduced the distribution in 1809 as a way of rationalizing the method of least squares as outlined above. Among English speakers, both "normal distribution" and "Gaussian distribution" are in common use, with different terms preferred by different communities.

Markov Chain

A Markov chain (discrete-time Markov chain or DTMC) named after Andrey Markov, is a mathematical system that undergoes transitions from one state to another, between a finite or countable number of possible states. It is a random process usually characterized as memoryless: the next state depends only on the current state and not on the sequence of events that preceded it. This specific kind of "memorylessness" is called the Markov property. Markov chains have many applications as statistical models of real-world processes.

Formally, a Markov chain is a random process with the Markov property. Often, the term "Markov chain" is used to mean a Markov process which has a discrete (finite or countable) state-space. Usually a Markov chain is defined for a discrete set of times (i.e., a discrete-time Markov chain) although some authors use the same terminology where "time" can take continuous values. The use of the term in Markov chain Monte Carlo methodology covers cases where the process is in discrete time (discrete algorithm steps) with a continuous state space. The following concentrates on the discrete-time discrete-state-space case.

A discrete-time random process involves a system which is in a certain state at each step, with the state changing randomly between steps. The steps are often thought of as moments in time, but they can equally well refer to physical distance or any other discrete measurement; formally, the steps are the integers or natural numbers, and the random process is a mapping of these to states. The Markov property states that the conditional probability distribution for the system at the next step (and in fact at all future steps) depends only on the current state of the system, and not additionally on the state of the system at previous steps.

Since the system changes randomly, it is generally impossible to predict with certainty the state of a Markov chain at a given point in the future. However, the statistical properties of the system's future can be predicted. In many applications, it is these statistical properties that are important.

The changes of state of the system are called transitions, and the probabilities associated with various state-changes are called transition probabilities. The set of all states and transition probabilities completely characterizes a Markov chain. By convention, we assume all possible states and transitions have been included in the definition of the processes, so there is always a next state and the process goes on forever.

A famous Markov chain is the so-called "drunkard's walk", a random walk on the number line where, at each step, the position may change by +1 or –1 with equal probability. From any position there are two possible transitions, to the next or previous integer. The transition probabilities depend only on the current position, not on the manner in which the position was reached. For example, the transition probabilities from 5 to 4 and 5 to 6 are both 0.5, and all other transition probabilities from 5 are 0. These probabilities are independent of whether the system was previously in 4 or 6.

Another example is the dietary habits of a creature who eats only grapes, cheese, or lettuce, and whose dietary habits conform to the following rules:

- It eats exactly once a day.
- If it ate cheese today, tomorrow it will eat lettuce or grapes with equal probability.
- If it ate grapes today, tomorrow it will eat grapes with probability 1/10, cheese with probability 4/10 and lettuce with probability 5/10.
- If it ate lettuce today, it will not eat lettuce again tomorrow but will eat grapes with probability 4/10 or cheese with probability 6/10.

This creature's eating habits can be modelled with a Markov chain since its choice tomorrow depends solely on what it ate today, not what it ate yesterday or even farther in the past. One statistical property that could be calculated is the expected percentage, over a long period, of the days on which the creature will eat grapes.

A series of independent events (for example, a series of coin flips) satisfies the formal definition of a Markov chain. However, the theory is usually applied only when the probability distribution of the next step depends non-trivially on the current state. Many other examples of Markov chains exist.

Formal Definition

A Markov chain is a sequence of random variablesX_1, X_2, X_3, ... with the Markov property, namely that, given the present state, the future and past states are independent. Formally,

$$\Pr(X_{n+1} = x \mid X_1 = x_1, X_2 = x_2, \ldots, X_n = x_n) = \Pr(X_{n+1} = x \mid X_n = x_n).$$

The possible values of X_i form a countable set S called the state space of the chain.

Markov chains are often described by a directed graph, where the edges are labelled by the probabilities of going from one state to the other states.

Variations

- Continuous-time Markov processes have a continuous index.
- Time-homogeneous Markov chains (or stationary Markov chains) are processes where

  $$\Pr(X_{n+1} = x \mid X_n = y) = \Pr(X_n = x \mid X_{n-1} = y)$$

  for all n. The probability of the transition is independent of n.
- A Markov chain of order m (or a Markov chain with memory m), where m is finite, is a process satisfying

  $$\begin{aligned} &\Pr(X_n = x_n \mid X_{n-1} = x_{n-1}, X_{n-2} = x_{n-2}, \ldots, X_1 = x_1) \\ =\ &\Pr(X_n = x_n \mid X_{n-1} = x_{n-1}, X_{n-2} = x_{n-2}, \ldots, X_{n-m} = x_{n-m}) \text{ for } n > m \end{aligned}$$

  In other words, the future state depends on the past m states. It is possible to construct a chain (Y_n) from (X_n) which has the 'classical' Markov property by taking as state space the ordered m-tuples of X values, ie. $Y_n = (X_n, X_{n-1}, \ldots, X_{n-m+1})$.
- An additive Markov chain of order m is determined by an additive conditional probability,

  $$\Pr(X_n = x_n \mid X_{n-1} = x_{n-1}, X_{n-2} = x_{n-2}, \ldots, X_{n-m} = x_{n-m}) = \sum_{r=1}^{m} f(x_n, x_{n-r}, r).$$

  The value $f(x_n, x_{n-r}, r)$ is the additive contribution of the variable x_{n-r} to the conditional probability.

Example

A simple example is shown in the figure below, using a directed graph to picture the state transitions. The states represent whether the economy is in a bull market, a bear market, or a recession, during a given week. According to the figure, a bull week is followed by another bull week 90% of the time, a bear market 7.5% of the time, and a recession the other 2.5%. Labelling the state space {1=bull, 2=bear, 3=recession} the transition matrix for this example is

$$P = \begin{bmatrix} 0.9 & 0.075 & 0.025 \\ 0.15 & 0.8 & 0.05 \\ 0.25 & 0.25 & 0.5 \end{bmatrix}.$$

The distribution over states can be written as a stochastic row vector x with the relation $x^{(n+1)} = x^{(n)}P$. So if at time n the system is in state

2=bear then 3 time periods later at time $n+3$ the distribution is

$$\begin{aligned}
x^{(n+3)} &= x^{(n+2)}P = \left(x^{(n+1)}P\right)P \\
&= x^{(n+1)}P^2 = \left(x^{(n)}P^2\right)P \\
&= x^{(n)}P^3 \\
&= \begin{bmatrix} 0 & 1 & 0 \end{bmatrix} \begin{bmatrix} 0.9 & 0.075 & 0.025 \\ 0.15 & 0.8 & 0.05 \\ 0.25 & 0.25 & 0.5 \end{bmatrix}^3 \\
&= \begin{bmatrix} 0 & 1 & 0 \end{bmatrix} \begin{bmatrix} 0.7745 & 0.17875 & 0.04675 \\ 0.3575 & 0.56825 & 0.07425 \\ 0.4675 & 0.37125 & 0.16125 \end{bmatrix} \\
&= \begin{bmatrix} 0.3575 & 0.56825 & 0.07425 \end{bmatrix}.
\end{aligned}$$

From this figure it is possible to calculate, for example, the long-term fraction of time during which the economy is in a recession, or on average how long it will take to go from a recession to a bull market. Using the transition probabilities, the steady-state probabilities indicate that 62.5% of weeks will be in a bull market, 31.25% of weeks will be in a bear market and 6.25% of weeks will be in a recession.

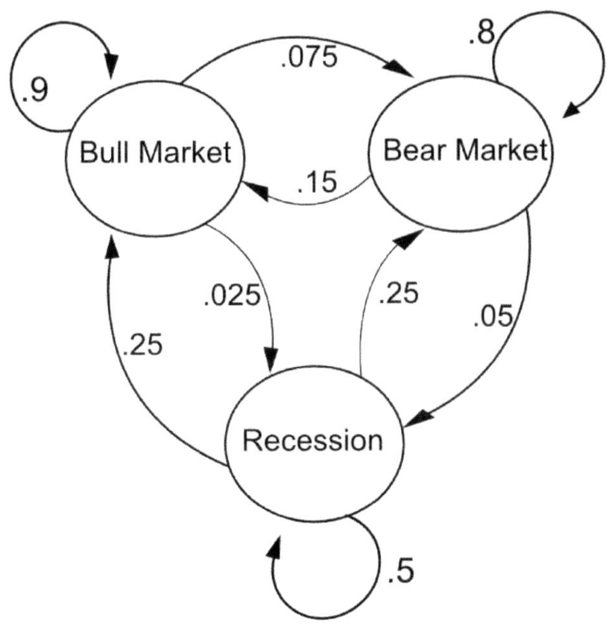

A thorough development and many examples can be found in the on-line monograph Meyn & Tweedie 2005.

A finite state machine can be used as a representation of a Markov chain. Assuming a sequence of independent and identically distributed input signals (for example, symbols from a binary alphabet chosen by coin tosses), if the machine is in state y at time n, then the probability that it moves to state x at time $n+1$ depends only on the current state.

Reducibility

A state j is said to be accessible from a state i (written $i \to j$) if a system started in state i has a non-zero probability of transitioning into state j at some point. Formally, state j is accessible from state i if there exists an integer $n \geq 0$ such that

$$\Pr(X_n = j \mid X_0 = i) = p_{ij}^{(n)} > 0.$$

Allowing n to be zero means that every state is defined to be accessible from itself.

A state i is said to communicate with state j (written $i \leftrightarrow j$) if both $i \to j$ and $j \to i$. A set of states C is a communicating class if every pair of states in C communicates with each other, and no state in C communicates with any state not in C. It can be shown that communication in this sense is an equivalence relation and thus that communicating classes are the equivalence classes of this relation. A communicating class is closed if the probability of leaving the class is zero, namely that if i is in C but j is not, then j is not accessible from i.

A state i is said to be essential or final if for all j such that $i \to j$ it is also true that $j \to i$. A state i is in essential if it is not essential.

Finally, a Markov chain is said to be irreducible if its state space is a single communicating class; in other words, if it is possible to get to any state from any state.

Periodicity

A state i has period k if any return to state i must occur in multiples of k time steps. Formally, the period of a state is defined as

$$k = \gcd\{n : \Pr(X_n = i \mid X_0 = i) > 0\}$$

(where "gcd" is the greatest common divisor). Note that even though a state has period k, it may not be possible to reach the state in k steps. For example, suppose it is possible to return to the state in

{6, 8, 10, 12, ...} time steps; k would be 2, even though 2 does not appear in this list.

If $k = 1$, then the state is said to be aperiodic: returns to state i can occur at irregular times. In other words, a state i is aperiodic if there exists n such that for all $n' \geq n$,

$$\Pr(X_{n'} = i \mid X_0 = i) > 0.$$

Otherwise ($k > 1$), the state is said to be periodic with period k. A Markov chain is aperiodic if every state is aperiodic.

An irreducible markov chain only needs one aperiodic state to imply all states are aperiodic.

Every state of a bipartite graph has an even period.

Recurrence

A state i is said to be transient if, given that we start in state i, there is a non-zero probability that we will never return to i. Formally, let the random variable T_i be the first return time to state i (the "hitting time"):

$$T_i = \inf\{n \geq 1 : X_n = i \mid X_0 = i\}.$$

The number

$$f_{ii}^{(n)} = \Pr(T_i = n)$$

is the probability that we return to state i for the first time after n steps. Therefore, state i is transient if

$$\Pr(T_i < \infty) = \sum_{n=1}^{\infty} f_{ii}^{(n)} < 1.$$

State i is recurrent (or persistent) if it is not transient. Recurrent states have finite hitting time with probability *1*.

Mean Recurrence Time

Even if the hitting time is finite with probability *1*, it need not have a finite expectation.

The mean recurrence time at state i is the expected return time M_i:

$$M_i = E[T_i] = \sum_{n=1}^{\infty} n \cdot f_{ii}^{(n)}.$$

State i is positive recurrent (or non-null persistent) if M_i is finite; otherwise, state i is null recurrent (or null persistent).

Expected Number of Visits

It can be shown that a state i is recurrent if and only if the expected number of visits to this state is infinite, i.e.,

$$\sum_{n=0}^{\infty} p_{ii}^{(n)} = \infty.$$

Absorbing states

A state i is called absorbing if it is impossible to leave this state. Therefore, the state i is absorbing if and only if

$$p_{ii} = 1 \text{ and } p_{ij} = 0 \text{ for } i \neq j.$$

If every state can reach an absorbing state, then the Markov chain is an absorbing Markov chain.

Ergodicity

A state i is said to be ergodic if it is aperiodic and positive recurrent. In other words, a state i is ergodic if it is recurrent, has a period of *1* and it has finite mean recurrence time. If all states in an irreducible Markov chain are ergodic, then the chain is said to be ergodic.

It can be shown that a finite state irreducible Markov chain is ergodic if it has an aperiodic state. A model has the ergodic property if there's a finite number N such that any state can be reached from any other state in exactly N steps. In case of a fully connected transition matrix where all transitions have a non-zero probability, this condition is fulfilled with N=1. A model with more than one state and just one out-going transition per state cannot be ergodic.

Steady-state Analysis and Limiting Distributions

If the Markov chain is a time-homogeneous Markov chain, so that the process is described by a single, time-independent matrix p_{ij}, then the vector π is called a stationary distribution (or invariant measure) if its entries π_j are non-negative and sum to 1 and if it satisfies

$$\pi_j = \sum_{i \in S} \pi_i p_{ij}.$$

An irreducible chain has a stationary distribution if and only if all of its states are positive recurrent. In that case, π is unique and is related to the expected return time:

$$\pi_j = \frac{C}{M_j},$$

where C is the normalizing constant. Further, if the chain is both irreducible and aperiodic, then for any i and j,

$$\lim_{n\to\infty} p_{ij}^{(n)} = \frac{C}{M_j}.$$

Note that there is no assumption on the starting distribution; the chain converges to the stationary distribution regardless of where it begins. Such π is called the equilibrium distribution of the chain.

If a chain has more than one closed communicating class, its stationary distributions will not be unique (consider any closed communicating class C_i in the chain; each one will have its own unique stationary distribution π_i.

Extending these distributions to the overall chain, setting all values to zero outside the communication class, yields that the set of invariant measures of the original chain is the set of all convex combinations of the π_i's). However, if a state j is aperiodic, then

$$\lim_{n\to\infty} p_{jj}^{(n)} = \frac{C}{M_j}$$

and for any other state i, let f_{ij} be the probability that the chain ever visits state j if it starts at i,

$$\lim_{n\to\infty} p_{ij}^{(n)} = C\frac{f_{ij}}{M_j}.$$

If a state i is periodic with period $k > 1$ then the limit

$$\lim_{n\to\infty} p_{ii}^{(n)}$$

does not exist, although the limit

$$\lim_{n\to\infty} p_{ii}^{(kn+r)}$$

does exist for every integer r.

Steady-state Analysis and the Time-inhomogeneous Markov Chain

A Markov chain need not necessarily be time-homogeneous to have an equilibrium distribution. If there is a probability distribution over states π such that

$$\pi_j = \sum_{i\in S} \pi_i \Pr(X_{n+1} = j \quad X_n = i)$$

for every state j and every time n then π is an equilibrium distribution of the Markov chain. Such can occur in Markov chain Monte Carlo (MCMC) methods in situations where a number of different transition matrices are used, because each is efficient for a particular kind of mixing, but each matrix respects a shared equilibrium distribution.

Finite state Space

If the state space is finite, the transition probability distribution can be represented by a matrix, called the transition matrix, with the (i, j)th element of P equal to

$$p_{ij} = \Pr(X_{n+1} = j \ X_n = i).$$

Since each row of P sums to one and all elements are non-negative, P is a right stochastic matrix.

Time-homogeneous Markov Chain with a Finite State Space

If the Markov chain is time-homogeneous, then the transition matrix P is the same after each step, so the k-step transition probability can be computed as the k-th power of the transition matrix, P^k.

The stationary distribution π is a (row) vector, whose entries are non-negative and sum to 1, that satisfies the equation

$$\pi = \pi \mathrm{P}.$$

In other words, the stationary distribution π is a normalized (meaning that the sum of its entries is 1) left eigenvector of the transition matrix associated with the eigenvalue 1.

Alternatively, π can be viewed as a fixed point of the linear (hence continuous) transformation on the unit simplex associated to the matrix P. As any continuous transformation in the unit simplex has a fixed point, a stationary distribution always exists, but is not guaranteed to be unique, in general. However, if the Markov chain is irreducible and aperiodic, then there is a unique stationary distribution π. Additionally, in this case P^k converges to a rank-one matrix in which each row is the stationary distribution π, that is,

$$\lim_{k\to\infty} P^k = 1\pi$$

where 1 is the column vector with all entries equal to 1. This is stated by the Perron–Frobenius theorem. If, by whatever means, $\lim_{k\to\infty} P^k$ is found, then the stationary distribution of the Markov chain in question can be easily determined for any starting distribution, as will be explained below.

For some stochastic matrices P, the limit $\lim_{k\to\infty} P^k$ does not exist, as shown by this example:

$$\begin{pmatrix} 0 & 1 \\ 1 & 0 \end{pmatrix}.$$

Because there are a number of different special cases to consider, the process of finding this limit if it exists can be a lengthy task. However, there are many techniques that can assist in finding this limit. Let P be an $n\times n$ matrix, and define $Q = \lim_{k\to\infty} P^k$.

It is always true that

$$QP = Q.$$

Subtracting Q from both sides and factoring then yields

$$Q(P - I_n) = 0_{n,n},$$

where I_n is the identity matrix of size n, and $0_{n,n}$ is the zero matrix of size $n\times n$. Multiplying together stochastic matrices always yields another stochastic matrix, so Q must be a stochastic matrix. It is sometimes sufficient to use the matrix equation above and the fact that Q is a stochastic matrix to solve for Q. Including the fact that the sum of each the rows in P is 1, there are $n+1$ equations for determining n unknowns, so it is computationally easier if on the one hand one selects one row in Q and substitute each of its elements by one, and on the other one substitute the corresponding element (the one in the same column) in the vector 0, and next left-multiply this latter vector by the inverse of transformed former matrix to find Q.

Here is one method for doing so: first, define the function $f(A)$ to return the matrix A with its right-most column replaced with all 1's. If $[f(P - I_n)]^{-1}$ exists then

$$Q = f(0_{n,n})[f(P - I_n)]^{-1}.$$

Explain: The original matrix equation is equivalent to a system of $n\times n$ linear equations in $n\times n$ variables. And there are n more linear equations from the fact that Q is a right stochastic matrix whose each row sums to 1. So it needs any $n\times n$ independent linear equations of the $(n\times n+n)$ equations to solve for the $n\times n$ variables. In this example, the n equations from "Q multiplied by the right-most column of (P-In)" have been replaced by the n stochastic ones.

One thing to notice is that if P has an element $P_{i,i}$ on its main diagonal that is equal to 1 and the ith row or column is otherwise filled with 0's, then that row or column will remain unchanged in all of the subsequent powers P^k. Hence, the ith row or column of Q will have the 1 and the 0's in the same positions as in P.

Convergence Speed to the Stationary Distribution

As stated earlier, from the equation $\pi = \pi P$, (if exists) the stationary (or stead state) distribution π is a) left eigenvector of row stochastic matrix P. Then assuming that P is diagonalizable or equivalently that P has n linearly independent eigenvectors, speed of convergence is elaborated as follows. For non-diagonalizable matrices, one may start with "Jordan Canonical Form" (*almost* diagonal form) of P and proceed with a bit more involved set of arguments in a similar way.

Let U be the matrix of eigenvectors (each normalized to having an L2 norm equal to 1) where each column is a left eigenvector of P and let Σ be the diagonal matrix of left eigenvalues of P, i.e. $\Sigma = \text{diag}(\lambda_1, \lambda_2, \lambda_3, \ldots, \lambda_n)$. Then by eigendecomposition

$$P = U\Sigma U^{-1}.$$

Let the eigenvalues be enumerated such that $1 = |\lambda_1| > |\lambda_2| \geq |\lambda_3| \geq \ldots \geq |\lambda_n|$. Since P is a row stochastic matrix, its largest left eigenvalue is 1. If there is a unique stationary distribution, then the largest eigenvalue and the corresponding eigenvector is unique too (because there is no other π which solves the stationary distribution equation above). Let u_i be the ith column of U matrix, i.e. u_i is the left eigenvector of P corresponding to λ_i. Also let x be an arbitrary length n row vector in the span of the eigenvectors u_i, that is

$$x^T = \sum_{i=1}^{n} a_i u_i$$

for some set of $a_i \in \mathbb{R}$. If we start multiplying P with x from left and continue this operation with the results, in the end we get the stationary distribution π. In other words $\pi = u_i \leftarrow xPPP \ldots P = xP^k$ as k goes to infinity. That means

$$\pi^{(k)} = x(U\Sigma U^{-1})(U\Sigma U^{-1})\ldots(U\Sigma U^{-1})$$

$$= xU\Sigma^k U^{-1}$$

since $UU^{-1} = \text{I}$ the identity matrix and power of a diagonal matrix is also a diagonal matrix where each entry is taken to that power.

$$= (a_1 \text{u}_1^T + a_2 \text{u}_2^T + \ldots + a_n \text{u}_n^T)\text{U}\Sigma^k \text{U}^{-1},$$

$$= a_1 \lambda_1^k \text{u}_1 + a_2 \lambda_2^k \text{u}_2 + \ldots + a_n \lambda_n^k \text{u}_n,$$

since the eigenvectors are orthonormal. Then

$$= \lambda_1^k \left\{ a_1 \mathrm{u}_1 + a_2 \left(\frac{\lambda_2}{\lambda_1} \right)^k \mathrm{u}_2 + a_3 \left(\frac{\lambda_3}{\lambda_1} \right)^k \mathrm{u}_3 + \ldots + a_n \left(\frac{\lambda_n}{\lambda_1} \right)^k \mathrm{u}_n \right\}.$$

Since $\pi = \mathrm{u}_1$, $\pi^{(k)}$ approaches to π as k goes to infinity with a speed in the order of λ_2/λ_1 exponentially. This follows because $|\lambda_2| \geq |\lambda_3| \geq \ldots \geq |\lambda_n|$, hence λ_2/λ_1 is the dominant term. Random noise in the state distribution π can also speed up this convergence to the stationary distribution.

Reversible Markov Chain

A Markov chain is said to be reversible if there is a probability distribution over states, π, such that

$$\pi_i \Pr(X_{n+1} = j \mid X_n = i) = \pi_j \Pr(X_{n+1} = i \mid X_n = j)$$

for all times n and all states i and j. This condition is also known as the detailed balance condition (some books refer the local balance equation). With a time-homogeneous Markov chain, $\Pr(X_{n+1} = j \mid X_n = i)$ does not change with time n and it can be written more simply as p_{ij}. In this case, the detailed balance equation can be written more compactly as

$$\pi_i p_{ij} = \pi_j p_{ji}.$$

Summing the original equation over i gives

$$\begin{aligned} \sum_i \pi_i \Pr(X_{n+1} = j \mid X_n = i) &= \sum_i \pi_j \Pr(X_{n+1} = i \mid X_n = j) \\ &= \pi_j \sum_i \Pr(X_{n+1} = i \mid X_n = j) = \pi_j, \end{aligned}$$

so, for reversible Markov chains, π is always a steady-state distribution of $\Pr(X_{n+1} = j \mid X_n = i)$ for every n.

If the Markov chain begins in the steady-state distribution, *i.e.*, if $\Pr(X_0 = i) = \pi_i$, then $\Pr(X_n = i) = \pi_i$ for all n and the detailed balance equation can be written as

$$\Pr(X_n = i, X_{n+1} = j) = \Pr(X_{n+1} = i, X_n = j).$$

The left- and right-hand sides of this last equation are identical except for a reversing of the time indices n and $n + 1$.

Reversible Markov chains are common in Markov chain Monte Carlo (MCMC) approaches because the detailed balance equation for a desired distribution π necessarily implies that the Markov chain has been constructed so that π is a steady-state distribution. Even with time-inhomogeneous Markov chains, where multiple transition matrices are used, if each such transition matrix exhibits detailed balance with the desired π distribution, this necessarily implies that π is a steady-state distribution of the Markov chain.

Bernoulli Scheme

A Bernoulli scheme is a special case of a Markov chain where the transition probability matrix has identical rows, which means that the next state is even independent of the current state (in addition to being independent of the past states). A Bernoulli scheme with only two possible states is known as a Bernoulli process.

General State Space

Many results for Markov chains with finite state space can be generalized to chains with uncountable state space through Harris chains. The main idea is to see if there is a point in the state space that the chain hits with probability one. Generally, it is not true for continuous state space, however, we can define sets A and B along with a positive number ε and a probability measure ρ, such that

If $\tau_A = \inf\{n \geq 0 : X_n \in A\}$, then $P_z(\tau_A < \infty) > 0$ for all z.

If $x \in A$ and $C \subset B$, then $p(x,C) \geq \varepsilon\rho(C)$.

Then we could collapse the sets into an auxiliary point α, and a recurrent Harris chain can be modified to contain α. Lastly, the collection of Harris chains is a comfortable level of generality, which is broad enough to contain a large number of interesting examples, yet restrictive enough to allow for a rich theory.

Chapter

7

Standard Deviation

In statistics and probability theory, standard deviation (represented by the symbol sigma, σ) shows how much variation or "dispersion" exists from the average (mean), or expected value. A low standard deviation indicates that the data points tend to be very close to the mean; high standard deviation indicates that the data points are spread out over a large range of values.

The standard deviation of a random variable, statistical population, data set, or probability distribution is the square root of its variance. It is algebraically simpler though practically less robust than the average absolute deviation. A useful property of standard deviation is that, unlike variance, it is expressed in the same units as the data. Note, however, that for measurements with percentage as unit, the standard deviation will have percentage points as unit.

In addition to expressing the variability of a population, standard deviation is commonly used to measure confidence in statistical conclusions. For example, the margin of error in polling data is determined by calculating the expected standard deviation in the results if the same poll were to be conducted multiple times. The reported margin of error is typically about twice the standard deviation – the radius of a 95 percent confidence interval. In science, researchers commonly report the standard deviation of experimental data, and only effects that fall far outside the range of standard deviation are considered statistically significant – normal random error or variation in the measurements is in this way distinguished from causal variation.

Standard deviation is also important in finance, where the standard deviation on the rate of return on an investment is a measure of the volatility of the investment.

When only a sample of data from a population is available, the standard deviation of the population can be estimated by a modified quantity called the "sample standard deviation". When the sample is the entire population, its unmodified standard deviation is called the "population standard deviation".

Consider a population consisting of the following eight values:

$$2, 4, 4, 4, 5, 5, 7, 9.$$

These eight data points have the mean (average) of 5:

$$\frac{2+4+4+4+5+5+7+9}{8} = 5.$$

To calculate the population standard deviation, first compute the difference of each data point from the mean, and square the result of each:

$$\begin{aligned} &(2-5)^2 = (-3)^2 = 9 && (5-5)^2 = 0^2 = 0 \\ &(4-5)^2 = (-1)^2 = 1 && (5-5)^2 = 0^2 = 0 \\ &(4-5)^2 = (-1)^2 = 1 && (7-5)^2 = 2^2 = 4 \\ &(4-5)^2 = (-1)^2 = 1 && (9-5)^2 = 4^2 = 16. \end{aligned}$$

Next, compute the average of these values, and take the square root:

$$\sqrt{\frac{(9+1+1+1+0+0+4+16)}{8}} = 2.$$

This quantity is the *population* standard deviation, and is equal to the square root of the variance.

The formula is valid only if the eight values we began with form the complete population. If the values instead were a random sample drawn from some larger parent population, then we would have divided by 7 (which is n–1) instead of 8 (which is n) in the denominator of the last formula, and then the quantity thus obtained would be called the *sample* standard deviation.

As a slightly more complicated real-life example, the average height for adult men in the United States is about 70 in, with a standard deviation of around 3 in. This means that most men (about 68 percent, assuming a normal distribution) have a height within 3 in of the mean (67–73 in) – one standard deviation – and almost all men (about 95%) have a height within 6 in of the mean (64–76 in) – two standard deviations. If the standard deviation were zero, then all men would be exactly 70 in tall. If the standard deviation were 20 in, then men would have much more variable heights, with a typical range of about 50–90 in. Three standard deviations account for 99.7 percent of the sample population being studied, assuming the distribution is normal (bell-shaped).

Definition of Population Values

Let X be a random variable with mean value μ:

$$E[X] = \mu.$$

Here the operator E denotes the average or expected value of X. Then the standard deviation of X is the quantity

$$\begin{aligned}\sigma &= \sqrt{\mathrm{E}[(X-\mu)^2]} \\ &= \sqrt{\mathrm{E}[X^2] + \mathrm{E}[(-2\mu X)] + \mathrm{E}[\mu^2]} = \sqrt{\mathrm{E}[X^2] - 2\mu\,\mathrm{E}[X] + \mu^2} = \sqrt{\mathrm{E}[X^2] - 2\mu^2 + \mu^2} = \sqrt{\mathrm{E}[X^2] - \mu^2} \\ &= \sqrt{\mathrm{E}[X^2] - (\mathrm{E}[X])^2}.\end{aligned}$$

(derived using the properties of expected value)

In other words the standard deviation σ (sigma) is the square root of the variance of X; i.e., it is the square root of the average value of $(X - \mu)^2$.

The standard deviation of a (univariate) probability distribution is the same as that of a random variable having that distribution. Not all random variables have a standard deviation, since these expected values need not exist. For example, the standard deviation of a random variable that follows a Cauchy distribution is undefined because its expected value μ is undefined.

Discrete Random Variable

In the case where X takes random values from a finite data set $x_1, x_2, \ldots, x_N$, with each value having the same probability, the standard deviation is

$$\sigma = \sqrt{\frac{1}{N}\left[(x_1-\mu)^2 + (x_2-\mu)^2 + \cdots + (x_N-\mu)^2\right]}, \text{ where } \mu = \frac{1}{N}(x_1 + \cdots + x_N),$$

or, using summation notation,

$$\sigma = \sqrt{\frac{1}{N}\sum_{i=1}^{N}(x_i-\mu)^2}, \text{ where } \mu = \frac{1}{N}\sum_{i=1}^{N}x_i.$$

If, instead of having equal probabilities, the values have different probabilities, let x_1 have **probability p_1, x_2 have probabil**ity $p_2, \ldots, x_N$

have probability p_N. In this case, the standard deviation will be

$$\sigma = \sqrt{\sum_{i=1}^{N} p_i (x_i - \mu)^2}, \text{ where } \mu = \sum_{i=1}^{N} p_i x_i.$$

Continuous Random Variable

The standard deviation of a continuous real-valued random variable X with probability density function $p(x)$ is

$$\sigma = \sqrt{\int_X (x-\mu)^2 \, p(x) dx}, \text{ where } \mu = \int_X x \, p(x) dx,$$

and where the integrals are definite integrals taken for x ranging over the set of possible values of the random variable X.

In the case of a parametric family of distributions, the standard deviation can be expressed in terms of the parameters. For example, in the case of the log-normal distribution with parameters μ and σ^2, the standard deviation is $[(\exp(\sigma^2) - 1)\exp(2\mu + \sigma^2)]^{1/2}$.

Estimation

One can find the standard deviation of an entire population in cases (such as standardized testing) where every member of a population is sampled. In cases where that cannot be done, the standard deviation σ is estimated by examining a random sample taken from the population. An estimator for σ used when sample size is very large is the standard deviation of the sample, denoted by s_N and defined as follows:

$$s_N = \sqrt{\frac{1}{N}\sum_{i=1}^{N} (x_i - \overline{x})^2}.$$

where $\{x_1, x_2, \ldots, x_N\}$ are the observed values of the sample items and $\overline{x}$ is the mean value of these observations, while the denominator N stands for the size of the sample. This estimator has a uniformly smaller mean squared error than the *sample standard deviation* below, and is the maximum-likelihood estimate when the population is normally distributed. But this estimator, when applied to a small or moderately sized sample, tends to be too low making it a biased estimator. The estimator for σ for small populations takes the general formulation above and applies Bessel's correction which uses degrees of freedom rather than sample size. Denoted by s, it is defined as follows:

$$s = \sqrt{\frac{1}{N-1}\sum_{i=1}^{N} (x_i - \overline{x})^2},$$

Where $N-1$ equals the number of degrees of freedom in the vector of residuals, $(x_1-\overline{x},\ldots,x_n-\overline{x})$. The estimator s^2 is unbiased for the variance σ^2 of the underlying population, if that variance exists and the sample values are drawn independently with replacement. However s is not an unbiased estimator of σ.

Confidence Interval of a Sampled Standard Deviation

The standard deviation we obtain by sampling a distribution is itself not absolutely accurate. This is especially true if the number of samples is very low. This effect can be described by the confidence interval or CI. For example, for N=2 the 95% CI of the SD is from 0.45*SD to 31.9*SD.

In other words the standard deviation of the distribution in 95% of the cases can be up to a factor of 31 larger or up to a factor 2 smaller. For N=10 the interval is 0.69*SD to 1.83*SD, the actual SD can still be almost a factor 2 higher than the sampled SD. For N=100 this is down to 0.88*SD to 1.16*SD.

So to be sure the sampled SD is close to the actual SD we need to sample a large number of points.

Identities and Mathematical Properties

The standard deviation is invariant under changes in location, and scales directly with the scale of the random variable. Thus, for a constant c and random variables X and Y:

$$\mathrm{stdev}(c)=0$$

$$\mathrm{stdev}(X+c)=\mathrm{stdev}(X),$$

$$\mathrm{stdev}(cX)=|c|\,\mathrm{stdev}(X).$$

The standard deviation of the sum of two random variables can be related to their individual standard deviations and the covariance between them:

$$\mathrm{stdev}(X+Y)=\sqrt{\mathrm{var}(X)+\mathrm{var}(Y)+2\mathrm{cov}(X,Y)}.$$

where $\mathrm{var}=\mathrm{stdev}^2$ and cov stand for variance and covariance, respectively.

The calculation of the sum of squared deviations can be related to moments calculated directly from the data. The standard deviation of the sample can be computed as:

$$\mathrm{stdev}(X)=\sqrt{E[(X-E(X))^2]}=\sqrt{E[X^2]-(E[X])^2}.$$

The sample standard deviation can be computed as:

$$\operatorname{stdev}(X) = \sqrt{\frac{N}{N-1}}\sqrt{E[(X - E(X))^2]}.$$

For a finite population with equal probabilities at all points, we have

$$\sqrt{\frac{1}{N}\sum_{i=1}^{N}(x_i - \overline{x})^2} = \sqrt{\frac{1}{N}\left(\sum_{i=1}^{N}x_i^2\right) - \overline{x}^2} = \sqrt{\frac{1}{N}\sum_{i=1}^{N}x_i^2 - \left(\frac{1}{N}\sum_{i=1}^{N}x_i\right)^2}.$$

This means that the standard deviation is equal to the square root of (the average of the squares less the square of the average).

Interpretation and Application

A large standard deviation indicates that the data points are far from the mean and a small standard deviation indicates that they are clustered closely around the mean.

For example, each of the three populations {0, 0, 14, 14}, {0, 6, 8, 14} and {6, 6, 8, 8} has a mean of 7. Their standard deviations are 7, 5, and 1, respectively. The third population has a much smaller standard deviation than the other two because its values are all close to 7. It will have the same units as the data points themselves. If, for instance, the data set {0, 6, 8, 14} represents the ages of a population of four siblings in years, the standard deviation is 5 years. As another example, the population {1000, 1006, 1008, 1014} may represent the distances travelled by four athletes, measured in meters. It has a mean of 1007 meters, and a standard deviation of 5 meters.

Standard deviation may serve as a measure of uncertainty. In physical science, for example, the reported standard deviation of a group of repeated measurements gives the precision of those measurements. When deciding whether measurements agree with a theoretical prediction, the standard deviation of those measurements is of crucial importance: if the mean of the measurements is too far away from the prediction (with the distance measured in standard deviations), then the theory being tested probably needs to be revised. This makes sense since they fall outside the range of values that could reasonably be expected to occur, if the prediction were correct and the standard deviation appropriately quantified.

While the standard deviation does measure how far typical values tend to be from the mean, other measures are available. An example is the mean absolute deviation, which might be considered a more direct measure of average distance, compared to the root mean square distance inherent in the standard deviation.

Application Examples

The practical value of understanding the standard deviation of a set of values is in appreciating how much variation there is from the average (mean).

Climate

As a simple example, consider the average daily maximum temperatures for two cities, one inland and one on the coast. It is helpful to understand that the range of daily maximum temperatures for cities near the coast is smaller than for cities inland. Thus, while these two cities may each have the same average maximum temperature, the standard deviation of the daily maximum temperature for the coastal city will be less than that of the inland city as, on any particular day, the actual maximum temperature is more likely to be farther from the average maximum temperature for the inland city than for the coastal one.

Particle Physics

Particle physics uses a standard of "5 sigma" for the declaration of a discovery. At five-sigma there is only one chance in nearly two million that a random fluctuation would yield the result. This level of certainty prompted the announcement that a particle consistent with the Higgs boson has been discovered in two independent experiments at CERN.

Sports

Another way of seeing it is to consider sports teams. In any set of categories, there will be teams that rate highly at some things and poorly at others. Chances are, the teams that lead in the standings will not show such disparity but will perform well in most categories. The lower the standard deviation of their ratings in each category, the more balanced and consistent they will tend to be. Teams with a higher standard deviation, however, will be more unpredictable. For example, a team that is consistently bad in most categories will have a low standard deviation. A team that is consistently good in most categories will also have a low standard deviation.

However, a team with a high standard deviation might be the type of team that scores a lot (strong offence) but also concedes a lot (weak defence), or, vice versa, that might have a poor offence but compensates by being difficult to score on. Trying to predict which teams, on any given day, will win, may include looking at the standard deviations of the various team "stats" ratings, in which anomalies can match strengths vs. weaknesses to attempt to understand what factors may prevail as stronger indicators of eventual scoring outcomes.

In racing, a driver is timed on successive laps. A driver with a low standard deviation of lap times is more consistent than a driver with a higher standard deviation. This information can be used to help understand where opportunities might be found to reduce lap times.

Finance

In finance, standard deviation is often used as a measure of the risk associated with price-fluctuations of a given asset (stocks, bonds, property, etc.), or the risk of a portfolio of assets (actively managed mutual funds, index mutual funds, or ETFs). Risk is an important factor in determining how to efficiently manage a portfolio of investments because it determines the variation in returns on the asset and/or portfolio and gives investors a mathematical basis for investment decisions (known as mean-variance optimization). The fundamental concept of risk is that as it increases, the expected return on an investment should increase as well, an increase known as the risk premium. In other words, investors should expect a higher return on an investment when that investment carries a higher level of risk or uncertainty. When evaluating investments, investors should estimate both the expected return and the uncertainty of future returns. Standard deviation provides a quantified estimate of the uncertainty of future returns.

For example, let's assume an investor had to choose between two stocks. Stock A over the past 20 years had an average return of 10 percent, with a standard deviation of 20 percentage points (pp) and Stock B, over the same period, had average returns of 12 percent but a higher standard deviation of 30 pp. On the basis of risk and return, an investor may decide that Stock A is the safer choice, because Stock B's additional two percentage points of return is not worth the additional 10 pp standard deviation (greater risk or uncertainty of the expected return). Stock B is likely to fall short of the initial investment (but also to exceed the initial investment) more often than Stock A under the same circumstances, and is estimated to return only two percent more on average. In this example, Stock A is expected to earn about 10 percent, plus or minus 20 pp (a range of 30 percent to "10 percent), about two-thirds of the future year returns. When considering more extreme possible returns or outcomes in future, an investor should expect results of as much as 10 percent plus or minus 60 pp, or a range from 70 percent to "50 percent, which includes outcomes for three standard deviations from the average return (about 99.7 percent of probable returns).

Calculating the average (or arithmetic mean) of the return of a security over a given period will generate the expected return of the asset. For each period, subtracting the expected return from the actual return results in the difference from the mean. Squaring the difference in each period and taking the average gives the overall variance of the return of the asset. The larger the variance, the greater risk the security carries. Finding the square root of this variance will give the standard deviation of the investment tool in question. Population standard deviation is used to set the width of Bollinger Bands, a widely adopted technical analysis tool. For example, the upper Bollinger Band is given as $x + n\sigma_x$. The most commonly used value for n is 2; there is about a five percent chance of going outside, assuming a normal distribution of returns.

Unfortunately, financial time series are known to be non-stationary series, whereas the statistical calculations above, such as standard deviation, apply only to stationary series. Whatever apparent "predictive powers" or "forecasting ability" that may appear when applied as above is illusory. To apply the above statistical tools to non-stationary series, the series first must be transformed to a stationary series, enabling use of statistical tools that now have a valid basis from which to work.

Geometric Interpretation

To gain some geometric insights and clarification, we will start with a population of three values, x_1, x_2, x_3. This defines a point $P = (x_1, x_2, x_3)$ in R^3. Consider the line $L = \{(r, r, r) : r \in \mathrm{R}\}$. This is the "main diagonal" going through the origin. If our three given values were all equal, then the standard deviation would be zero and P would lie on L. So it is not unreasonable to assume that the standard deviation is related to the *distance* of P to L. And that is indeed the case. To move orthogonally from L to the point P, one begins at the point:

$$M = (\overline{x}, \overline{x}, \overline{x})$$

whose coordinates are the mean of the values we started out with. A little algebra shows that the distance between P and M (which is the same as the orthogonal distance between P and the line L) is equal to the standard deviation of the vector x_1, x_2, x_3, multiplied by the square root of the number of dimensions of the vector (3 in this case.)

Relationship between Standard Deviation and Mean

The mean and the standard deviation of a set of data are descriptive statistics usually reported together. In a certain sense, the standard deviation is a "natural" measure of statistical dispersion if the centre of the data is measured about the mean. This is because the standard deviation from the mean is smaller than from any other point. The precise statement is the floowing suppose x_1, ..., x_n are real numbers and define the function:

$$\sigma(r) = \sqrt{\frac{1}{N-1}\sum_{i=1}^{N}(x_i - r)^2}.$$

Using calculus or by completing the square, it is possible to show that $\sigma(r)$ has a unique minimum at the mean:

$$r = \bar{x}.$$

Variability can also be measured by the coefficient of variation, which is the ratio of the standard deviation to the mean. It is a dimensionless number.

Standard Deviation of the Mean

Often, we want some information about the precision of the mean we obtained. We can obtain this by determining the standard deviation of the sampled mean. The standard deviation of the mean is related to the standard deviation of the distribution by:

$$\sigma_{\text{mean}} = \frac{1}{\sqrt{N}}\sigma$$

where N is the number of observations in the sample used to estimate the mean. This can easily be proven with :

$$\begin{aligned}
\text{var}(X) &\equiv \sigma_X^2 \\
\text{var}(X_1 + X_2) &\equiv \text{var}(X_1) + \text{var}(X_2) \\
\text{var}(cX_1) &\equiv c^2\,\text{var}(X_1)
\end{aligned}$$

hence

$$\begin{aligned}
\text{var}(\text{mean}) &= \text{var}\left(\frac{1}{N}\sum_{i=1}^{N} X_i\right) = \frac{1}{N^2}\text{var}\left(\sum_{i=1}^{N} X_i\right) \\
&= \frac{1}{N^2}\sum_{i=1}^{N}\text{var}(X_i) = \frac{N}{N^2}\text{var}(X) = \frac{1}{N}\text{var}(X).
\end{aligned}$$

Resulting in:

$$\sigma_{\text{mean}} = \frac{\sigma}{\sqrt{N}}.$$

Rapid Calculation Methods

The following two formulas can represent a running (continuous) standard deviation. A set of three power sums s_0, s_1, s_2 are each computed over a set of N values of x, denoted as x_1, ..., x_N:

$$s_j = \sum_{k=1}^{N} x_k^j.$$

Note that s_0 raises x to the zero power, and since x^0 is always 1, s_0 evaluates to N. Given the results of these three running summations, the values s_0, s_1, s_2 can be used at any time to compute the *current* value of the running standard deviation:

$$\sigma = \frac{\sqrt{s_0 s_2 - s_1^2}}{s_0}$$

Similarly for sample standard deviation,

$$s = \sqrt{\frac{s_0 s_2 - s_1^2}{s_0 (s_0 - 1)}}.$$

In a computer implementation, as the three s_j sums become large, we need to consider round-off error, arithmetic overflow, and arithmetic underflow. The method below calculates the running sums method with reduced rounding errors. This is a "one pass" algorithm for calculating variance of n samples without the need to store prior data during the calculation. Applying this method to a time series will result in successive values of standard deviation corresponding to n data points as n grows larger with each new sample, rather than a constant-width sliding window calculation.

For $k = 1, ..., n$:

$$A_0 = 0$$

$$A_k = A_{k-1} + \frac{x_k - A_{k-1}}{k}$$

where A is the mean value.

$$Q_0 = 0$$

$$Q_k = Q_{k-1} + \frac{k-1}{k}(x_k - A_{k-1})^2 = Q_{k-1} + (x_k - A_{k-1})(x_k - A_k)$$

Sample variance:

$$s_n^2 = \frac{Q_n}{n-1}$$

Population variance:

$$\sigma_n^2 = \frac{Q_n}{n}$$

Weighted Calculation

When the values x_i are weighted with unequal weights w_i, the power sums s_0, s_1, s_2 are each computed as:

$$s_j = \sum_{k=1}^{N} w_k x_k^j.$$

And the standard deviation equations remain unchanged. Note that s_0 is now the sum of the weights and not the number of samples N.

The incremental method with reduced rounding errors can also be applied, with some additional complexity.

A running sum of weights must be computed for each k from 1 to n:

$$\begin{aligned} W_0 &= 0 \\ W_k &= W_{k-1} + w_k \end{aligned}$$

and places where $1/n$ is used above must be replaced by w_i/W_n:

$$\begin{aligned} A_0 &= 0 \\ A_k &= A_{k-1} + \frac{w_k}{W_k}(x_k - A_{k-1}) \\ Q_0 &= 0 \\ Q_k &= Q_{k-1} + \frac{w_k W_{k-1}}{W_k}(x_k - A_{k-1})^2 = Q_{k-1} + w_k(x_k - A_{k-1})(x_k - A_k) \end{aligned}$$

In the final division,

$$\sigma_n^2 = \frac{Q_n}{W_n}$$

and

$$s_n^2 = \frac{n'}{n'-1}\sigma_n^2$$

where n is the total number of elements, and n' is the number of elements with non-zero weights. The above formulas become equal to the simpler formulas given above if weights are taken as equal to one.

Combining Standard Deviations

Population-based Statistics

The populations of sets, which may overlap, can be calculated simply as follows:

$$\begin{aligned} & N_{X\cup Y} &= N_X + N_Y - N_{X\cap Y} \\ X\cap Y = \varnothing \Rightarrow \ & N_{X\cap Y} &= 0 \\ \Rightarrow \ & N_{X\cup Y} &= N_X + N_Y \end{aligned}$$

Standard deviations of non-overlapping ($X \cap Y = \varnothing$) sub-populations can be aggregated as follows if the size (actual or relative to one another) and means of each are known:

$$\mu_{X\cup Y} = \frac{N_X\mu_X + N_Y\mu_Y}{N_X + N_Y}$$

$$\sigma_{X\cup Y} = \sqrt{\frac{N_X\sigma_X^2 + N_Y\sigma_Y^2}{N_X + N_Y} + \frac{N_X N_Y}{(N_X + N_Y)^2}(\mu_X - \mu_Y)^2}$$

For example, suppose it is known that the average American man has a mean height of 70 inches with a standard deviation of three inches and that the average American woman has a mean height of 65 inches with a standard deviation of two inches.

Also assume that the number of men, N, is equal to the number of women.

Then the mean and standard deviation of heights of American adults could be calculated as:

$$\mu = \frac{N\cdot 70 + N\cdot 65}{N + N} = \frac{70 + 65}{2} = 67.5$$

$$\sigma = \sqrt{\frac{3^2 + 2^2}{2} + \frac{(70 - 65)^2}{2^2}} = \sqrt{12.75} \approx 3.57$$

For the more general case of M non-overlapping populations, X_1 through X_M, and the aggregate population $X = \bigcup_i X_i$:

$$\mu_X = \frac{\sum_i N_{X_i}\mu_{X_i}}{\sum_i N_{X_i}}$$

$$\sigma_X = \sqrt{\frac{\sum_i N_{X_i}(\sigma_{X_i}^2 + \mu_{X_i}^2)}{\sum_i N_{X_i}} - \mu_X^2} = \sqrt{\frac{\sum_i N_{X_i}\sigma_{X_i}^2}{\sum_i N_{X_i}} + \frac{\sum_{i<j} N_{X_i}N_{X_j}(\mu_{X_i} - \mu_{X_j})^2}{(\sum_i N_{X_i})^2}}$$

where

$$X_i \cap X_j = \varnothing, \quad \forall\, i < j.$$

If the size (actual or relative to one another), mean, and standard deviation of two overlapping populations are known for the populations as well as their intersection, then the standard deviation of the overall population can still be calculated as follows:

$$\mu_{X\cup Y} = \frac{1}{N_{X\cup Y}}(N_X\mu_X + N_Y\mu_Y - N_{X\cap Y}\mu_{X\cap Y})$$

$$\sigma_{X\cup Y} = \sqrt{\frac{1}{N_{X\cup Y}}(N_X[\sigma_X^2 + \mu_X^2] + N_Y[\sigma_Y^2 + \mu_Y^2] - N_{X\cap Y}[\sigma_{X\cap Y}^2 + \mu_{X\cap Y}^2]) - \mu_{X\cup Y}^2}$$

If two or more sets of data are being added together datapoint by datapoint, the standard deviation of the result can be calculated if the standard deviation of each data set and the covariance between each pair of data sets is known:

$$\sigma_X = \sqrt{\sum_i \sigma_{X_i}^2 + \sum_{i,j} \operatorname{cov}(X_i, X_j)}$$

For the special case where no correlation exists between any pair of data sets, then the relation reduces to the root-mean-square:

$$\operatorname{cov}(X_i, X_j) = 0, \quad \forall i < j$$

$$\Rightarrow \quad \sigma_X = \sqrt{\sum_i \sigma_{X_i}^2}.$$

Sample-based Statistics

Standard deviations of non-overlapping ($X \cap Y = \varnothing$) sub-samples can be aggregated as follows if the actual size and means of each are known:

$$\mu_{X\cup Y} = \frac{1}{N_{X\cup Y}}(N_X\mu_X + N_Y\mu_Y)$$

$$\sigma_{X\cup Y} = \sqrt{\frac{1}{N_{X\cup Y} - 1}([N_X - 1]\sigma_X^2 + N_X\mu_X^2 + [N_Y - 1]\sigma_Y^2 + N_Y\mu_Y^2 - [N_X + N_Y]\mu_{X\cup Y}^2)}$$

For the more general case of M non-overlapping data sets, X_1 through X_M, and the aggregate data set $X = \bigcup_i X_i$:

$$\mu_X = \frac{1}{\sum_i N_{X_i}}\left(\sum_i N_{X_i}\mu_{X_i}\right)$$

$$\sigma_X = \sqrt{\frac{1}{\sum_i N_{X_i} - 1}\left(\sum_i \left[(N_{X_i} - 1)\sigma_{X_i}^2 + N_{X_i}\mu_{X_i}^2\right] - \left[\sum_i N_{X_i}\right]\mu_X^2\right)}$$

where:

$$X_i \cap X_j = \varnothing, \quad \forall i < j.$$

If the size, mean, and standard deviation of two overlapping samples are known for the samples as well as their intersection, then the standard deviation of the aggregated sample can still be calculated. In general:

$$\mu_{X \cup Y} = \frac{1}{N_{X \cup Y}}(N_X \mu_X + N_Y \mu_Y - N_{X \cap Y} \mu_{X \cap Y})$$

$$\sigma_{X \cup Y} = \sqrt{\frac{1}{N_{X \cup Y} - 1}([N_X - 1]\sigma_X^2 + N_X \mu_X^2 + [N_Y - 1]\sigma_Y^2 + N_Y \mu_Y^2 - [N_{X \cap Y} - 1]\sigma_{X \cap Y}^2 - N_{X \cap Y} \mu_{X \cap Y}^2 - [N_X + N_Y - N_{X \cap Y}]\mu_{X \cup Y}^2)}$$

History

The term *standard deviation* was first used in writing by Karl Pearson in 1894, following his use of it in lectures. This was as a replacement for earlier alternative names for the same idea: for example, Gauss used *mean error*. It may be worth noting in passing that the mean error is mathematically distinct from the standard deviation.

Variance

In probability theory and statistics, the variance is a measure of how far a set of numbers is spread out. It is one of several descriptors of a probability distribution, describing how far the numbers lie from the mean (expected value). In particular, the variance is one of the moments of a distribution. In that context, it forms part of a systematic approach to distinguishing between probability distributions. While other such approaches have been developed, those based on moments are advantageous in terms of mathematical and computational simplicity.

The variance is a parameter describing in part either the actual probability distribution of an observed population of numbers, or the theoretical probability distribution of a sample (a not-fully-observed population) of numbers. In the latter case a sample of data from such a distribution can be used to construct an estimate of its variance: in the simplest cases this estimate can be the sample variance.

Basic Discussion

Examples: The variance of a random variable or distribution is the expectation, or mean, of the squared deviation of that variable from its expected value or mean. Thus the variance is a measure of the amount of variation of the values of that variable, taking account of all possible values and their probabilities or weightings (not just the extremes which give the range). For example, a perfect six-sided die, when thrown, has expected value of

$$\frac{1}{6}(1+2+3+4+5+6)=3.5.$$

Its expected absolute deviation—the mean of the equally likely absolute deviations from the mean—is

$$\frac{1}{6}(|1-3.5|+|2-3.5|+|3-3.5|+|4-3.5|+|5-3.5|+|6-3.5|)$$

$$=\frac{1}{6}(2.5+1.5+0.5+0.5+1.5+2.5)=1.5.$$

But its expected *squared* deviation—its variance (the mean of the equally likely squared deviations)—is

$$\frac{1}{6}(2.5^2+1.5^2+0.5^2+0.5^2+1.5^2+2.5^2)=17.5/6\approx 2.9.$$

As another example, if a coin is tossed twice, the number of heads is: 0 with probability 0.25, 1 with probability 0.5 and 2 with probability 0.25. Thus the expected value of the number of heads is:

$$0.25\times 0+0.5\times 1+0.25\times 2=1,$$

and the variance is:

$$0.25\times(0-1)^2+0.5\times(1-1)^2+0.25\times(2-1)^2=0.25+0+0.25=0.5.$$

Units of Measurement

Unlike expected absolute deviation, the variance of a variable has units that are the square of the units of the variable itself. For example, a variable measured in inches will have a variance measured in square inches.

For this reason, describing data sets via their standard deviation or root mean square deviation is often preferred over using the variance. In the dice example the standard deviation is $\sqrt{2.9}\approx 1.7$, slightly larger than the expected absolute deviation of 1.5.

The standard deviation and the expected absolute deviation can both be used as an indicator of the "spread" of a distribution. The standard deviation is more amenable to algebraic manipulation than the expected absolute deviation, and, together with variance and its generalization covariance, is used frequently in theoretical statistics; however the expected absolute deviation tends to be more robust as it is less sensitive to outliers arising from measurement anomalies or an unduly heavy-tailed distribution.

Estimating the Variance

Real-world distributions such as the distribution of yesterday's rain throughout the day are typically not fully known, unlike the behaviour of perfect dice or an ideal distribution such as the normal distribution, because it is impractical to account for every raindrop. Instead one estimates the mean and variance of the whole distribution as the computed mean and variance of a sample of n observations drawn suitably randomly from the whole sample space, in this example the set of all measurements of yesterday's rainfall in all available rain gauges.

This method of estimation is close to optimal, with the caveat that it underestimates the variance by a factor of $(n - 1) / n$. (For example, when $n = 1$ the variance of a single observation is obviously zero regardless of the true variance). This gives a bias which should be corrected for when n is small by multiplying by $n / (n - 1)$. If the mean is determined in some other way than from the same samples used to estimate the variance then this bias does not arise and the variance can safely be estimated as that of the samples.

To illustrate the relation between the population variance and the sample variance, suppose that in the (not entirely observed) population of numerical values, the value 1 occurs 1/3 of the time, the value 2 occurs 1/3 of the time, and the value 4 occurs 1/3 of the time. The population mean is (1/3)[1 + 2 + 4] = 7/3. The equally likely deviations from the population mean are 1– 7/3, 2 – 7/3, and 4 – 7/3. The population variance — the expected squared deviation from the mean 7/3 — is $(1/3)[(-4/3)^2 + (-1/3)^2 + (5/3)^2] = 14/9$.

Now suppose for the sake of a simple example that we take a very small sample of $n = 2$ observations, and consider the nine equally likely possibilities for the set of numbers within that sample:

(1, 1), (1, 2), (1,4), (2, 1), (2,2), (2, 4), (4,1), (4, 2), and (4, 4). For these nine possible samples, the sample variance of the two numbers is respectively 0, 1/4, 9/4, 1/4, 0, 4/4, 9/4, 4/4, and 0. With our plan to observe two values, we could end up computing any of these sample variances (and indeed if we hypothetically could observe a pair of numbers many times, we would compute each of these sample variances 1/9 of the time). So the expected value, over all possible samples that might be drawn from the population, of the computed sample variance is (1/9)[0 + 1/4 + 9/4 + 1/4 + 0 + 4/4 + 9/4 + 4/4 + 0] = 7/9. This value of 7/9 for the expected value of our sample variance computation is a substantial underestimate of the true population variance, which we computed as 14/9, because our sample size of just two observations was so small. But if we adjust for this downward bias by multiplying our computed sample variance, whichever it may be, by $n/(n-1) = 2/(2-1) = 2$, then our estimate of the population variance would be any one of 0, 1/2, 9/2, 1/2, 0, 4/2, 9/2, 4/2, and 0. The average of these is indeed the correct population variance of 14/9, so on average over all possible samples we would have the correct estimate of the population variance.

The variance of a real-valued random variable is its second central moment, and it also happens to be its second cumulant. Just as some distributions do not have a mean, some do not have a variance. The mean exists whenever the variance exists, but the converse is not necessarily true.

Definition

If a random variable X has the expected value (mean) $\mu = \mathrm{E}[X]$, then the variance of X is the covariance of X with itself, given by:

$$\begin{aligned}\mathrm{Var}(X) &= \mathrm{Cov}(X,X)\\ &= \mathrm{E}[(X-\mu)(X-\mu)]\\ &= \mathrm{E}\left[(X-\mu)^2\right].\end{aligned}$$

That is, the variance is the expected value of the squared difference between the variable's realization and the variable's mean. This definition encompasses random variables that are discrete, continuous, neither, or mixed. From the corresponding expression for Covariance, it can be expanded:

$$\begin{aligned}\mathrm{Var}(X) &= \mathrm{Cov}(X,X)\\ &= \mathrm{E}[XX]-\mathrm{E}[X]\mathrm{E}[X]\\ &= \mathrm{E}\left[X^2\right]-(\mathrm{E}[X])^2.\end{aligned}$$

A mnemonic for the above expression is "mean of square minus square of mean". The variance of random variable X is typically designated as Var(X), σ_X^2, or simply σ^2 (pronounced "sigma squared").

Continuous Random Variable

If the random variable X is continuous with probability density function $f(x)$, then the variance equals the second central moment, given by

$$\mathrm{Var}(X) = \sigma^2 = \int (x-\mu)^2 f(x)dx = \int x^2 f(x)dx - \mu^2$$

where μ is the expected value,

$$\mu = \int x f(x) dx$$

and where the integrals are definite integrals taken for *x* ranging over the range of *X*.

If a continuous distribution does not have an expected value, as is the case for the Cauchy distribution, it does not have a variance either.

Many other distributions for which the expected value does exist also do not have a finite variance because the integral in the variance definition diverges. An example is a Pareto distribution whose index k satisfies $1 < k \leq 2$.

Discrete Random Variable

If the random variable *X* is discrete with probability mass function $x_1 \mapsto p_1, \ldots, x_n \mapsto p_n$, then

$$\mathrm{Var}(X) = \sum_{i=1}^{n} (p_i \cdot (x_i - \mu)^2) = \sum_{i=1}^{n} (p_i \cdot x_i^2) - \mu^2$$

where μ is the expected value, i.e.

$$\mu = \sum_{i=1}^{n} p_i \cdot x_i .$$

(When such a discrete weighted variance is specified by weights whose sum is not 1, then one divides by the sum of the weights.)

That is, it is the expected value of the square of the deviation of *X* from its own mean. In plain language, it can be expressed as "The mean of the squares of the deviations of the data points from the average". It is thus the *mean squared deviation.*

Examples

Exponential Distribution: The exponential distribution with parameter λ is a continuous distribution whose support is the semi-infinite interval [0,∞). Its probability density function is given by:

$$f(x) = \lambda e^{-\lambda x},$$

and it has expected value $\mu = \lambda^{-1}$. Therefore the variance is equal to:

$$\int f(x)(x-\mu)^2 dx = \int_0^\infty \lambda e^{-\lambda x}(x-\lambda^{-1})^2 dx = \lambda^{-2}.$$

So for an exponentially distributed random variable $\sigma^2 = \mu^2$.

Fair Die

A six-sided fair die can be modelled with a discrete random variable with outcomes 1 through 6, each with equal probability $\frac{1}{6}$. The expected value is (1 + 2 + 3 + 4 + 5 + 6)/6 = 3.5. Therefore the variance can be computed to be:

$$\sum_{i=1}^{6}\tfrac{1}{6}(i-3.5)^2 = \tfrac{1}{6}\sum_{i=1}^{6}(i-3.5)^2 = \tfrac{1}{6}\left((-2.5)^2+(-1.5)^2+(-0.5)^2+0.5^2+1.5^2+2.5^2\right)$$
$$= \tfrac{1}{6}\cdot 17.50 = \tfrac{35}{12} \approx 2.92.$$

The general formula for the variance of the outcome X of a die of n sides is:

$$\sigma^2 = E(X^2)-(E(X))^2 = \frac{1}{n}\sum_{i=1}^{n} i^2 - \left(\frac{1}{n}\sum_{i=1}^{n} i\right)^2$$
$$= \tfrac{1}{6}(n+1)(2n+1) - \tfrac{1}{4}(n+1)^2$$
$$= \frac{n^2-1}{12}.$$

Properties

Basic Properties: Variance is non-negative because the squares are positive or zero.

$$\mathrm{Var}(X) \geq 0.$$

The variance of a constant random variable is zero, and if the variance of a variable in a data set is 0, then all the entries have the same value.

$$P(X=a)=1 \Leftrightarrow \mathrm{Var}(X)=0.$$

Variance is invariant with respect to changes in a location parameter. That is, if a constant is added to all values of the variable, the variance is unchanged.

$$\mathrm{Var}(X+a)=\mathrm{Var}(X).$$

If all values are scaled by a constant, the variance is scaled by the square of that constant.

$$\mathrm{Var}(aX)=a^2\,\mathrm{Var}(X).$$

The variance of a sum of two random variables is given by:

$$\mathrm{Var}(aX+bY)=a^2\,\mathrm{Var}(X)+b^2\,\mathrm{Var}(Y)+2ab\mathrm{Cov}(X,Y),$$
$$\mathrm{Var}(X-Y)=\mathrm{Var}(X)+\mathrm{Var}(Y)-2\mathrm{Cov}(X,Y),$$

In general we have for the sum of N random variables:

$$\operatorname{Var}\left(\sum_{i=1}^{N} X_i\right) = \sum_{i,j=1}^{N} \operatorname{Cov}(X_i, X_j) = \sum_{i=1}^{N} \operatorname{Var}(X_i) + \sum_{i \neq j} \operatorname{Cov}(X_i, X_j).$$

These results lead to the variance of a linear combination as:

$$\begin{aligned}\operatorname{Var}\left(\sum_{i=1}^{N} a_i X_i\right) &= \sum_{i=1}^{N}\sum_{j=1}^{N} a_i a_j \operatorname{Cov}(X_i, X_j) \\ &= \sum_{i=1}^{N} a_i^2 \operatorname{Var}(X_i) + \sum_{i \neq j} a_i a_j \operatorname{Cov}(X_i, X_j) \\ &= \sum_{i=1}^{N} a_i^2 \operatorname{Var}(X_i) + 2 \sum_{1 \le i < j \le N} a_i a_j \operatorname{Cov}(X_i, X_j).\end{aligned}$$

The variance of a finite sum of *uncorrelated* random variables is equal to the sum of their variances. This stems from the above identity and the fact that for uncorrelated variables the covariance is zero; that is, if

$$\operatorname{Cov}(X_i, X_j) = 0 \ (i \neq j),$$

then

$$\operatorname{Var}\left(\sum_{i=1}^{N} X_i\right) = \sum_{i=1}^{N} \operatorname{Var}(X_i).$$

Sum of Uncorrelated Variables (Bienaymé Formula)

One reason for the use of the variance in preference to other measures of dispersion is that the variance of the sum (or the difference) of uncorrelated random variables is the sum of their variances:

$$\operatorname{Var}\left(\sum_{i=1}^{n} X_i\right) = \sum_{i=1}^{n} \operatorname{Var}(X_i).$$

This statement is called the Bienaymé formula and was discovered in 1853. It is often made with the stronger condition that the variables are independent, but uncorrelatedness suffices. So if all the variables have the same variance σ^2, then, since division by n is a linear transformation, this formula immediately implies that the variance of their mean is

$$\operatorname{Var}(\bar{X}) = \operatorname{Var}\left(\frac{1}{n}\sum_{i=1}^{n} X_i\right) = \frac{1}{n^2}\sum_{i=1}^{n} \operatorname{Var}(X_i) = \frac{\sigma^2}{n}.$$

That is, the variance of the mean decreases when n increases.

This formula for the variance of the mean is used in the definition of the standard error of the sample mean, which is used in the central limit theorem.

Product of independent variables

If two variables X and Y are independent, the variance of their product is given by

$$\operatorname{Var}(XY) = [E(X)]^2 \operatorname{Var}(Y) + [E(Y)]^2 \operatorname{Var}(X) + \operatorname{Var}(X)\operatorname{Var}(Y).$$

Sum of Correlated Variables

In general, if the variables are correlated, then the variance of their sum is the sum of their covariances:

$$\operatorname{Var}\left(\sum_{i=1}^{n} X_i\right) = \sum_{i=1}^{n}\sum_{j=1}^{n} \operatorname{Cov}(X_i, X_j).$$

(Note: This by definition includes the variance of each variable, since $\operatorname{Cov}(X_i,X_i) = \operatorname{Var}(X_i)$.)

Here Cov is the covariance, which is zero for independent random variables (if it exists). The formula states that the variance of a sum is equal to the sum of all elements in the covariance matrix of the components. This formula is used in the theory of Cronbach's alpha in classical test theory.

So if the variables have equal variance σ^2 and the average correlation of distinct variables is ρ, then the variance of their mean is

$$\operatorname{Var}(\bar{X}) = \frac{\sigma^2}{n} + \frac{n-1}{n}\rho\sigma^2.$$

This implies that the variance of the mean increases with the average of the correlations. Moreover, if the variables have unit variance, for example if they are standardized, then this simplifies to

$$\operatorname{Var}(\bar{X}) = \frac{1}{n} + \frac{n-1}{n}\rho.$$

This formula is used in the Spearman–Brown prediction formula of classical test theory. This converges to ρ if n goes to infinity, provided that the average correlation remains constant or converges too.

So for the variance of the mean of standardized variables with equal correlations or converging average correlation we have

$$\lim_{n\to\infty} \operatorname{Var}(\bar{X}) = \rho.$$

Therefore, the variance of the mean of a large number of standardized variables is approximately equal to their average correlation.

This makes clear that the sample mean of correlated variables does generally not converge to the population mean, even though the Law of large numbers states that the sample mean will converge for independent variables.

Weighted Sum of Variables

The scaling property and the Bienaymé formula, along with this property from the covariance page: $\mathrm{Cov}(aX, bY) = ab\,\mathrm{Cov}(X, Y)$ jointly imply that

$$\mathrm{Var}(aX + bY) = a^2\,\mathrm{Var}(X) + b^2\,\mathrm{Var}(Y) + 2ab\mathrm{Cov}(X,Y).$$

This implies that in a weighted sum of variables, the variable with the largest weight will have a disproportionally large weight in the variance of the total. For example, if X and Y are uncorrelated and the weight of X is two times the weight of Y, then the weight of the variance of X will be four times the weight of the variance of Y.

The expression above can be extended to a weighted sum of multiple variables:

$$\mathrm{Var}\left(\sum_{i}^{n} a_i X_i\right) = \sum_{i=1}^{n} a_i^2\,\mathrm{Var}(X_i) + 2\sum_{1\le i}\sum_{<j\le n} a_i a_j\,\mathrm{Cov}(X_i, X_j)$$

Formulae for the Variance

The formula most often used for deriving the variance of a theoretical distribution is as follows:

$$\begin{aligned}\mathrm{Var}(X) &= \mathrm{E}(X^2 - 2X\mathrm{E}(X) + (\mathrm{E}(X))^2)\\ &= \mathrm{E}(X^2) - 2(\mathrm{E}(X))^2 + (\mathrm{E}(X))^2\\ &= \mathrm{E}(X^2) - (\mathrm{E}(X))^2.\end{aligned}$$

This will be useful when it is possible to derive formulae for the expected value and for the expected value of the square.

This formula is also sometimes used in connection with the sample variance. While useful for hand calculations, it is not advised for computer calculations as it suffers from catastrophic cancellation if the two components of the equation are similar in magnitude and floating point arithmetic is used. This is discussed below.

Calculation from the CDF

The population variance for a non-negative random variable can be expressed in terms of the cumulative distribution function F using

$$2\int_0^\infty uH(u)du - \left(\int_0^\infty H(u)du\right)^2.$$

where $H(u) = 1 - F(u)$ is the right tail function. This expression can be used to calculate the variance in situations where the CDF, but not the density, can be conveniently expressed.

Characteristic Property

The second moment of a random variable attains the minimum value when taken around the first moment (i.e., mean) of the random variable, i.e. $\arg\min_m E((X-m)^2) = E(X)$. Conversely, if a continuous function φ satisfies $\arg\min_m E(\varphi(X-m)) = E(X)$ for all random variables X, then it is necessarily of the form $\varphi(x) = ax^2 + b$, where $a > 0$. This also holds in the multidimensional case.

Matrix Notation for the Variance of a Linear Combination

Let's define X as a column vector of n random variables $X_1,\ldots,X_n$, and c as a column vector of N scalars $c_1,\ldots,c_n$.

Therefore $c^T X$ is a linear combination of these random variables, where c^T denotes the transpose of vector c. Let also be Σ the variance-covariance matrix of the vector X. The variance of $c^T X$ is given by:

$$\operatorname{Var}(c^T X) = c^T \Sigma c.$$

Approximating the Variance of a Function

The delta method uses second-order Taylor expansions to approximate the variance of a function of one or more random variables. For example, the approximate variance of a function of one variable is given by

$$\operatorname{Var}[f(X)] \approx \left(f'(\mathrm{E}[X])\right)^2 \operatorname{Var}[X]$$

provided that f is twice differentiable and that the mean and variance of X are finite.

Population Variance and Sample Variance

In general, the *population variance* of a *finite* population of size N is given by

$$\sigma^2 = \frac{1}{N}\sum_{i=1}^{N}(x_i - \mu)^2 = \left(\frac{1}{N}\sum_{i=1}^{N}x_i^2\right) - \mu^2$$

where

$$\mu = \frac{1}{N}\sum_{i=1}^{N}x_i$$

is the population mean, and

$$\begin{aligned}\sum_{i=1}^{N}(x_i - \mu)^2 &= \sum_{i=1}^{N}\left(x_i^2 - 2x_i\mu + \mu^2\right) \\ &= \sum_{i=1}^{N}\left(x_i^2 + \mu^2\right) - 2\mu\sum_{i=1}^{N}x_i \\ &= \sum_{i=1}^{N}\left(x_i^2 + \mu^2\right) - 2N\mu^2 \\ &= \sum_{i=1}^{N}\left(x_i^2 + \mu^2 - 2\mu^2\right) \\ &= \sum_{i=1}^{N}\left(x_i^2 - \mu^2\right)\end{aligned}$$

In many practical situations, the true variance of a population is not known *a priori* and must be computed somehow. When dealing with extremely large populations, it is not possible to count every object in the population. A common task is to estimate the variance of a population from a sample. We take a sample with replacement of n values $y_1, \ldots, y_n$ from the population, where $n < N$, and estimate the variance on the basis of this sample. There are several good estimators. Two of them are well known:

$$s_n^2 = \frac{1}{n}\sum_{i=1}^{n}(y_i - \overline{y})^2 = \left(\frac{1}{n}\sum_{i=1}^{n}y_i^2\right) - \overline{y}^2,$$

and

$$\begin{aligned}s^2 &= \frac{1}{n-1}\sum_{i=1}^{n}(y_i - \overline{y})^2 \\ &= \frac{1}{n-1}\sum_{i=1}^{n}\left(y_i^2 - \overline{y}^2\right) \\ &= \frac{1}{n-1}\sum_{i=1}^{n}y_i^2 - \frac{n}{n-1}\overline{y}^2.\end{aligned}$$

The first estimator, also known as the second central moment, is called the *biased sample variance*. The second estimator is called the *unbiased sample variance*. Either estimator may be simply referred to as the *sample variance* when the version can be determined by context. Here, $\bar{y}$ denotes the sample mean:

$$\bar{y} = \frac{1}{n}\sum_{i=1}^{n} y_i.$$

The two estimators only differ slightly as can be seen, and for larger values of the sample size n the difference is negligible. While the first one may be seen as the variance of the sample considered as a population, the second one is the unbiased estimator of the population variance, meaning that its expected value $E[s^2]$ is equal to the true variance of the sampled random variable; the use of the term $n-1$ is called Bessel's correction. In particular,

$$E[s^2] = \sigma^2,$$

while, in contrast,

$$E[s_n^2] = \frac{n-1}{n}\sigma^2.$$

The unbiased sample variance is a U-statistic for the function $f(x_1, x_2) = (x_1 - x_2)^2/2$, meaning that it is obtained by averaging a 2-sample statistic over 2-element subsets of the population.

Distribution of the Sample Variance

Being a function of random variables, the sample variance is itself a random variable, and it is natural to study its distribution. In the case that y_i are independent observations from a normal distribution, Cochran's theorem shows that s^2 follows a scaled chi-squared distribution:

$$(n-1)\frac{s^2}{\sigma^2} \sim \chi^2_{n-1}.$$

As a direct consequence, it follows that

$$E(s^2) = E\left(\frac{\sigma^2}{n-1}\chi^2_{n-1}\right) = \sigma^2,$$

and

$$\operatorname{Var}[s^2] = \operatorname{Var}\left(\frac{\sigma^2}{n-1}\chi^2_{n-1}\right) = \frac{\sigma^4}{(n-1)^2}\operatorname{Var}\left(\chi^2_{n-1}\right) = \frac{2\sigma^4}{n-1}.$$

If the y_i are independent and identically distributed, but not necessarily normally distributed, then

$$E[s^2] = \sigma^2, \quad Var[s^2] = \sigma^4 \left(\frac{2}{n-1} + \frac{\kappa}{n} \right) = \frac{1}{n} \left(\mu_4 - \frac{n-3}{n-1} \sigma^4 \right),$$

where κ is the excess kurtosis of the distribution and μ_4 is the fourth moment about the mean.

If the conditions of the law of large numbers hold for the squared observations, s^2 is a consistent estimator of σ^2.. One can see indeed that the variance of the estimator tends asymptotically to zero.

Samuelson's Inequality

Samuelson's inequality is a result that states, given that the sample mean and variance have been calculated from a particular sample, bounds on the values that individual values in the sample can take. Values must lie within the limits $m \pm s(n - 1)^{1/2}$.

Relations with the Harmonic and Arithmetic Means

It has been shown that for a sample of real numbers that

$$\text{Var} \leq 2M(A - H)$$

where M is the maximum of the sample, A is the arithmetic mean, H is the harmonic mean of the sample and *Var* is the variance of the sample. This bound has been improved on and it is known that variance is bounded by

$$\text{Var} \leq \frac{M(A-H)(M-A)}{M-H}$$

$$\text{Var} \geq \frac{m(A-H)(A-m)}{H-m}$$

where m is the minimum of the sample.

Generalizations

If $X$is a vector-valued random variable, with values in $\mathbb{R}^n$, and thought of as a column vector, then the natural generalization of variance is $E((X-\mu)(X-\mu)^T)$, where $\mu = E(X)$ and X^T is the transpose of X, and so is a row vector. This variance is a positive semi-definite square matrix, commonly referred to as the covariance matrix.

If X is a complex-valued random variable, with values in $\mathbb{C}$, then its variance is $E((X-\mu)(X-\mu)^\dagger)$, where $X^\dagger$ is the conjugate transpose of X. This variance is also a positive semi-definite square matrix.

Tests of Equality of Variances

Testing for the equality of two or more variances is difficult. The F test and chi square tests are both sensitive to non normality and are not recommended for this purpose.

Several non parametric tests have been proposed: these include the Barton-David-Ansari-Fruend-Siegel-Tukey test, the Capon test, Mood test, the Klotz test and the Sukhatme test. The Sukhatme test applies to two variances and requires that both medians be known and equal to zero. The Mood, Klotz, Capon and Barton-David-Ansari-Fruend-Siegel-Tukey tests also apply to two variances. They allow the median to be unknown but do require that the two medians are equal.

The Lehman test is a parametric test of two variances. Of this test there are several variants known. Other tests of the equality of variances include the Box test, the Box-Anderson test and the Moses test. Resampling methods, which include the bootstrap and the jackknife, may be used to test the equality of variances.

History

The term *variance* was first introduced by Ronald Fisher in his 1918 paper *The Correlation Between Relatives on the Supposition of Mendelian Inheritance*:

The great body of available statistics show us that the deviations of a human measurement from its mean follow very closely the Normal Law of Errors, and, therefore, that the variability may be uniformly measured by the standard deviation corresponding to the square root of the mean square error.

When there are two independent causes of variability capable of producing in an otherwise uniform population distributions with standard deviations θ_1 and θ_2, it is found that the distribution, when both causes act together, has a standard deviation $\sqrt{\theta_1^2 + \theta_2^2}$. It is therefore desirable in analysing the causes of variability to deal with the square of the standard deviation as the measure of variability. We shall term this quantity the Variance...

Moment of Inertia

The variance of a probability distribution is analogous to the moment of inertia in classical mechanics of a corresponding mass distribution along a line, with respect to rotation about its centre of mass. It is because of this analogy that such things as the variance are called *moments* of probability distributions. The covariance matrix is related to the moment of inertia tensor for multivariate distributions. The moment of inertia of a cloud of n points with a covariance matrix of Σ is given by

$$I = n(1_{3\times3}\,\mathrm{tr}(\Sigma) - \Sigma).$$

This difference between moment of inertia in physics and in statistics is clear for points that are gathered along a line. Suppose many points are close to the x axis and distributed along it. The covariance matrix might look like

$$\Sigma = \begin{bmatrix} 10 & 0 & 0 \\ 0 & 0.1 & 0 \\ 0 & 0 & 0.1 \end{bmatrix}.$$

That is, there is the most variance in the x direction. However, physicists would consider this to have a low moment *about* the x axis so the moment-of-inertia tensor is

$$I = n\begin{bmatrix} 0.2 & 0 & 0 \\ 0 & 10.1 & 0 \\ 0 & 0 & 10.1 \end{bmatrix}.$$

Chapter

8

Rule of Sum

In combinatorics, the rule of sum or addition principle is a basic counting principle. Stated simply, it is the idea that if we have a ways of doing something and b ways of doing another thing and we can not do both at the same time, then there are $a + b$ ways to choose one of the actions.

More formally, the rule of sum is a fact about set theory. It states that sum of the sizes of a finite collection of pairwise disjoint sets is the size of the union of these sets. That is, if $S_1, S_2, ..., S_n$ are pairwise disjoint sets, then we have:

$$|S_1| + |S_2| + \cdots + |S_n| = |S_1 \cup S_2 \cup \cdots \cup S_n|$$

A woman has decided to shop at one store today, either in the north part of town or the south part of town. If she visits the north part of town, she will either shop at a mall, a furniture store, or a jewelry store (3 ways).

If she visits the south part of town then she will either shop at a clothing store or a shoe store (2 ways). Thus there are 3+2=5 possible shops the woman could end up shopping at today.

Inclusion-exclusion Principle

The inclusion-exclusion principle can be thought of as a generalization of the rule of sum in that it too enumerates the number of elements in the union of some sets (but does not require the sets to be disjoint).

It states that if $A_1, ..., A_n$ are finite sets, then

$$\left|\bigcup_{i=1}^{n} A_i\right| = \sum_{i=1}^{n}|A_i| - \sum_{i,j:1\le i<j\le n}|A_i \cap A_j| + \sum_{i,j,k:1\le i<j<k\le n}|A_i \cap A_j \cap A_k| - \cdots + (-1)^{n-1}|A_1 \cap \cdots \cap A_n|.$$

Probability

Probability (or likelihood) is a measure or estimation of how likely it is that something will happen or that a statement is true. Probabilities are given a value between 0 (0% chance or *will not happen*) and 1 (100% chance or *will happen*). The higher the degree of probability, the more likely the event is to happen, or, in a longer series of samples, the greater the number of times such event is expected to happen.

These concepts have been given an axiomatic mathematical derivation in probability theory, which is used widely in such areas of study as mathematics, statistics, finance, gambling, science, artificial intelligence/machine learning and philosophy to, for example, draw inferences about the expected frequency of events. Probability theory is also used to describe the underlying mechanics and regularities of complex systems.

Probability Interpretations

The word probability has been used in a variety of ways since it was first applied to the mathematical study of games of chance. Does probability measure the real, physical tendency of something to occur or is it a measure of how strongly one believes it will occur, or does it draw on both these elements? In answering such questions, we interpret the probability values of probability theory.

There are two broad categories of probability interpretations which can be called "physical" and "evidential" probabilities. Physical probabilities, which are also called objective or frequency probabilities, are associated with random physical systems such as roulette wheels, rolling dice and radioactive atoms. In such systems, a given type of event (such as the dice yielding a six) tends to occur at a persistent rate, or "relative frequency", in a long run of trials. Physical probabilities either explain, or are invoked to explain, these stable frequencies. Thus talk about physical probability makes sense only when dealing with well definedrandom experiments.

The two main kinds of theory of physical probability are frequentist accounts (such as those of Venn, Reichenbach and von Mises) and propensity accounts (such as those of Popper, Miller, Giere and Fetzer). Evidential probability, also called Bayesian probability (or subjectivist probability), can be assigned to any statement whatsoever, even when no random process is involved, as a way to represent its subjective plausibility, or the degree to which the statement is supported by the available evidence.

On most accounts, evidential probabilities are considered to be degrees of belief, defined in terms of dispositions to gamble at certain odds. The four main evidential interpretations are the classical (e.g. Laplace's) interpretation, the subjective interpretation (de Finetti and Savage), the epistemic or inductive interpretation (Ramsey, Cox) and the logical interpretation (Keynes and Carnap).

Some interpretations of probability are associated with approaches to statistical inference, including theories of estimation and hypothesis testing. The physical interpretation, for example, is taken by followers of "frequentist" statistical methods, such as R. A. Fisher, Jerzy Neyman and Egon Pearson. Statisticians of the opposing Bayesian school typically accept the existence and importance of physical probabilities, but also consider the calculation of evidential probabilities to be both valid and necessary in statistics. This article, however, focuses on the interpretations of probability rather than theories of statistical inference.

The terminology of this topic is rather confusing, in part because probabilities are studied within a variety of academic fields. The word "frequentist" is especially tricky. To philosophers it refers to a particular theory of physical probability, one that has more or less been abandoned. To scientists, on the other hand, "frequentist probability" is just what philosophers call physical (or objective) probability. Those who promote Bayesian inference view "frequentist statistics" as an approach to statistical inference that recognises only physical probabilities. Also the word "objective", as applied to probability, sometimes means exactly what "physical" means here, but is also used of evidential probabilities that are fixed by rational constraints, such as logical and epistemic probabilities.

It is unanimously agreed that statistics depends somehow on probability. But, as to what probability is and how it is connected with statistics, there has seldom been such complete disagreement and breakdown of communication since the Tower of Babel. Doubtless, much of the disagreement is merely terminological and would disappear under sufficiently sharp analysis.

Philosophy

The philosophy of probability presents problems chiefly in matters of epistemology and the uneasy interface between mathematical concepts and ordinary language as it is used by non-mathematicians. Probability theory is an established field of study in mathematics. It has its origins in correspondence discussing the mathematics of games of chance between Blaise Pascal and Pierre de Fermat in the seventeenth century, and was formalized and rendered axiomatic as a distinct branch of mathematics by Andrey Kolmogorov in the twentieth century. In its axiomatic form, mathematical statements about probability theory carry the same sort of epistemological confidence shared by other mathematical statements in the philosophy of mathematics.

The mathematical analysis originated in observations of the behaviour of game equipment such as playing cards and dice, which are designed specifically to introduce random and equalized elements; in mathematical terms, they are subjects of indifference. This is not the only way probabilistic statements are used in ordinary human language: when people say that "*it will probably rain*", they typically do not mean that the outcome of rain versus not-rain is a random factor that the odds currently favour; instead, such statements are perhaps better understood as qualifying their expectation of rain with a degree of confidence. Likewise, when it is written that "the most probable explanation" of the name of Ludlow, Massachusetts "is that it was named after Roger Ludlow", what is meant here is not that Roger Ludlow is favoured by a random factor, but rather that this is the most plausible explanation of the evidence, which admits other, less likely explanations.

Thomas Bayes attempted to provide a logic that could handle varying degrees of confidence; as such, Bayesian probability is an attempt to recast the representation of probabilistic statements as an expression of the degree of confidence by which the beliefs they express are held. Though probability initially had somewhat mundane motivations, its modern influence and use is widespread ranging from Evidence based medicine, through Six sigma, all the way to the Probabilistically checkable proof and the String theory landscape.

Classical Definition

The first attempt at mathematical rigour in the field of probability, championed by Pierre-Simon Laplace, is now known as the classical definition. Developed from studies of games of chance (such as rolling dice) it states that probability is shared equally between all the possible outcomes, provided these outcomes can be deemed equally likely.

The theory of chance consists in reducing all the events of the same kind to a certain number of cases equally possible, that is to say, to such as we may be equally undecided about in regard to their existence, and in determining the number of cases favourable to the event whose probability is sought. The ratio of this number to that of all the cases possible is the measure of this probability, which is thus simply a fraction whose numerator is the number of favourable cases and whose denominator is the number of all the cases possible.

– *Pierre-Simon Laplace, A Philosophical Essay on Probabilities*

This can be Represented Mathematically as follows: If a random experiment can result in N mutually exclusive and equally likely outcomes and if N_A of these outcomes result in the occurrence of the event *A*, the probability of *A* is defined by

$$P(A) = \frac{N_A}{N}$$

There are two clear limitations to the classical definition. Firstly, it is applicable only to situations in which there is only a 'finite' number of possible outcomes. But some important random experiments, such as tossing a coin until it rises heads, give rise to an infinite set of outcomes. And secondly, you need to determine in advance that all the possible outcomes are equally likely without relying on the notion of probability to avoid circularity—for instance, by symmetry considerations.

Frequentism

Figure: *For frequentists, the probability of the ball landing in any pocket can be determined only by repeated trials in which the observed result converges to the underlying probability* in the *long run.*

Frequentists posit that the probability of an event is its relative frequency over time, i.e., its relative frequency of occurrence after repeating a process a large number of times under similar conditions. This is also known as aleatory probability. The events are assumed to be governed by some random physical phenomena, which are either phenomena that are predictable, in principle, with sufficient information; or phenomena which are essentially unpredictable. Examples of the first kind include tossing dice or spinning a roulette wheel; an example of the second kind is radioactive decay. In the case of tossing a fair coin, frequentists say that the probability of getting a heads is 1/2, not because there are two equally likely outcomes but because repeated series of large numbers of trials demonstrate that the empirical frequency converges to the limit 1/2 as the number of trials goes to infinity.

If we denote by n_a the number of occurrences of an event $\mathcal{A}$ in n trials, then if $\lim_{n \to \infty} \frac{n_a}{n} = p$ we say that $P(\mathcal{A}) = p$

The frequentist view has its own problems. It is of course impossible to actually perform an infinity of repetitions of a random experiment to determine the probability of an event. But if only a finite number of repetitions of the process are performed, different relative frequencies will appear in different series of trials. If these relative frequencies are to define the probability, the probability will be slightly different every time it is measured. But the real probability should be the same every time. If we acknowledge the fact that we only can measure a probability with some error of measurement attached, we still get into problems as the error of measurement can only be expressed as a probability, the very concept we are trying to define. This renders even the frequency definition circular.

Logical, Epistemic, and Inductive Probability

It is widely recognised that the term "probability" is sometimes used in contexts where it has nothing to do with physical randomness. Consider, for example, the claim that the extinction of the dinosaurs was probably caused by a large meteorite hitting the earth. Statements such as "Hypothesis H is probably true" have been interpreted to mean that the (presently available) empirical evidence (*E*, say) supports *H* to a high degree. This degree of support of *H* by *E* has been called the logical probability of *H* given *E*, or the epistemic probability of *H* given *E*, or the inductive probability of *H* given *E*.

The differences between these interpretations are rather small, and may seem inconsequential. One of the main points of disagreement lies in the relation between probability and belief. Logical probabilities are conceived (for example in Keynes' Treatise on Probability) to be objective, logical relations between propositions (or sentences), and hence not to depend in any way upon belief. They are degrees of (partial) entailment, or degrees of logical consequence, not degrees of belief. (They do, nevertheless, dictate proper degrees of belief, as is discussed below.) Frank P. Ramsey, on the other hand, was skeptical about the existence of such objective logical relations and argued that (evidential) probability is "the logic of partial belief" ("Truth and Probability", 1926, p. 157). In other words, Ramsey held that epistemic probabilities simply *are* degrees of rational belief, rather than being logical relations that merely *constrain* degrees of rational belief.

Another point of disagreement concerns the *uniqueness* of evidential probability, relative to a given state of knowledge. Rudolf Carnap held, for example, that logical principles always determine a unique logical probability for any statement, relative to any body of evidence. Ramsey, by contrast, thought that while degrees of belief are subject to some rational constraints (such as, but not limited to, the axioms of probability) these constraints usually do not determine a unique value. Rational people, in other words, may differ somewhat in their degrees of belief, even if they all have the same information.

Propensity

Propensity theorists think of probability as a physical propensity, or disposition, or tendency of a given type of physical situation to yield an outcome of a certain kind or to yield a long run relative frequency of such an outcome. This kind of objective probability is sometimes called 'chance'.

Propensities, or chances, are not relative frequencies, but purported causes of the observed stable relative frequencies. Propensities are invoked to explain why repeating a certain kind of experiment will generate a given outcome type at a persistent rate, which are known as propensities or chances. Frequentists are unable to take this approach, since relative frequencies do not exist for single tosses of a coin, but only for large ensembles or collectives. In contrast, a propensitist is able to use the law of large numbers to explain the behaviour of long-run frequencies. This law, which is a consequence of the axioms of probability, says that if (for example) a coin is tossed repeatedly many times, in such a way that its probability of landing heads is the same on each toss, and the outcomes are probabilistically independent, then the relative frequency of heads will be close to the probability of heads on each single toss. This law allows that stable long-run frequencies are a manifestation of invariant *single-case* probabilities. In addition to explaining the emergence of stable relative frequencies, the idea of propensity is motivated by the desire to make sense of single-case probability attributions in quantum mechanics, such as the probability of decay of a particular atom at a particular time.

The main challenge facing propensity theories is to say exactly what propensity means. (And then, of course, to show that propensity thus defined has the required properties.) At present, unfortunately, none of the well-recognised accounts of propensity comes close to meeting this challenge.

A propensity theory of probability was given by Charles Sanders Peirce. A later propensity theory was proposed by philosopher Karl Popper, who had only slight acquaintance with the writings of C. S. Peirce, however. Popper noted that the outcome of a physical experiment is produced by a certain set of "generating conditions". When we repeat an experiment, as the saying goes, we really perform another experiment with a (more or less) similar set of generating conditions. To say that a set of generating conditions has propensity p of producing the outcome E means that those exact conditions, if repeated indefinitely, would produce an outcome sequence in which E occurred with limiting relative frequency p. For Popper then, a deterministic experiment would have propensity 0 or 1 for each outcome, since those generating conditions would have same outcome on each trial. In other words, non-trivial propensities (those that differ from 0 and 1) only exist for genuinely indeterministic experiments.

A number of other philosophers, including David Miller and Donald A. Gillies, have proposed propensity theories somewhat similar to Popper's.

Other propensity theorists (e.g. Ronald Giere) do not explicitly define propensities at all, but rather see propensity as defined by the theoretical role it plays in science. They argue, for example, that physical magnitudes such as electrical charge cannot be explicitly defined either, in terms of more basic things, but only in terms of what they do (such as attracting and repelling other electrical charges). In a similar way, propensity is whatever fills the various roles that physical probability plays in science.

What roles does physical probability play in science? What are its properties? One central property of chance is that, when known, it constrains rational belief to take the same numerical value. David Lewis called this the *Principal Principle*, a term that philosophers have mostly adopted. For example, suppose you are certain that a particular biased coin has propensity 0.32 to land heads every time it is tossed. What is then the correct price for a gamble that pays $1 if the coin lands heads, and nothing otherwise? According to the Principal Principle, the fair price is 32 cents.

Subjectivism

Subjectivists, also known as Bayesians or followers of epistemic probability, give the notion of probability a subjective status by regarding it as a measure of the 'degree of belief' of the individual assessing the uncertainty of a particular situation. Epistemic or subjective probability is sometimes called credence, as opposed to the term chance for a propensity probability. Some examples of epistemic probability are to assign a probability to the proposition that a proposed law of physics is true, and to determine how probable it is that a suspect committed a crime, based on the evidence presented. Gambling odds don't reflect the bookies' belief in a likely winner, so much as the other bettors' belief, because the bettors are actually betting against one another. The odds are set based on how many people have bet on a possible winner, so that even if the high odds players always win, the bookies will always make their percentages anyway.

The use of Bayesian probability raises the philosophical debate as to whether it can contribute valid justifications of belief. Bayesians point to the work of Ramsey and de Finetti as proving that subjective beliefs must follow the laws of probability if they are to be coherent. The use of Bayesian probability involves specifying a prior probability. This may be obtained from consideration of whether the required prior probability is greater or lesser than a reference probability associated with an urn model or a thought experiment. The issue is that for a given problem, multiple thought experiments could apply, and choosing one is a matter of judgement: different people may assign different prior probabilities, known as the reference class problem. The "sunrise problem" provides an example.

Practical Controversy

This difference in point of view has also many implications both for the methods by which statistics is practiced, and for the way in which conclusions are expressed. When comparing two hypotheses and using some information, frequency methods would typically result in the rejection or non-rejection of the original hypothesis at a particular significance level, and frequentists would all agree that the hypothesis should be rejected or not at that level of significance. However, there is no normative methodology to choose levels of significance. Bayesian methods would suggest that one hypothesis was more probable than the other, but individual Bayesians might differ about which was the more probable and by how much, by virtue of having used different priors; but that's the same thing as disagreeing on significance levels, except significance levels are just an *ad hoc* device which are not really a probability, while priors are not only justified by the rules of probability, but there is definitely a normative methodology to define beliefs; so even if a Bayesian wanted to express complete ignorance (as a frequentist claims to do but does it wrong), they could do it with the maximum entropy principle.

The most important distinction between the frequentist and Bayesian paradigms, is that frequentist makes strong distinctions between probability, statistics, and decision-making, whereas Bayesians unify decision-making, statistics and probability under a single philosophically and mathematically consistent framework, unlike the frequentist paradigm which has been proven to be inconsistent, especially for real-world situations where experiments (or "random events") can not be repeated more than once. Bayesians would argue that this is right and proper — if the issue is such that reasonable people can put forward different, but plausible, priors and the data are such that the likelihood does not swamp the prior, then the issue is not resolved unambiguously at the present stage of knowledge and Bayesian statistics highlights this fact. They would argue that any approach that purports to produce a single, definitive answer to the question at hand in these circumstances is obscuring the truth. But "frequentists" do not claim to produce "a single, definitive answer to the question at hand".

An alternative solution, is the eclectic view, which accepts both interpretations: depending on the situation, one selects one of the two interpretations for pragmatic, or principled, reasons.

Prediction

An alternative account of probability emphasizes the role of *prediction* – predicting future observations on the basis of past observations, not on unobservable parameters. In its modern form, it is mainly in the Bayesian vein. This was the main function of probability before the 20th century, but fell out of favour compared to the parametric approach, which modelled phenomena as a physical system that was observed with error, such as in celestial mechanics. The modern predictive approach was pioneered by Bruno de Finetti, with the central idea of exchangeability – that future observations should behave like past observations. This view came to the attention of the Anglophone world with the 1974 translation of de Finetti's book, and has since been propounded by such statisticians as Seymour Geisser.

Axiomatic Probability

The mathematics of probability can be developed on an entirely axiomatic basis that is independent of any interpretation.

Etymology

The word *Probability* derives from the Latin *probabilitas*, which can also mean *probity*, a measure of the authority of a witness in a legal case in Europe, and often correlated with the witness's nobility. In a sense, this differs much from the modern meaning of *probability*, which, in contrast, is a measure of the weight of empirical evidence, and is arrived at from inductive reasoning and statistical inference.

History

The scientific study of probability is a modern development. Gambling shows that there has been an interest in quantifying the ideas of probability for millennia, but exact mathematical descriptions arose much later. There are reasons of course, for the slow development of the mathematics of probability. Whereas games of chance provided the impetus for the mathematical study of probability, fundamental issues are still obscured by the superstitions of gamblers. According to Richard Jeffrey, "Before the middle of the seventeenth century, the term 'probable' (Latin *probabilis*) meant *approvable*, and was applied in that sense, univocally, to opinion and to action. A probable action or opinion was one such as sensible people would undertake or hold, in the circumstances." However, in legal contexts especially, 'probable' could also apply to propositions for which there was good evidence.

Aside from elementary work by Girolamo Cardano in the 16th century, the doctrine of probabilities dates to the correspondence of Pierre de Fermat and Blaise Pascal (1654). Christiaan Huygens (1657) gave the earliest known scientific treatment of the subject. Jakob Bernoulli's *Ars Conjectandi* (posthumous, 1713) and Abraham de

Moivre's *Doctrine of Chances* (1718) treated the subject as a branch of mathematics.

The theory of errors may be traced back to Roger Cotes's *Opera Miscellanea* (posthumous, 1722), but a memoir prepared by Thomas Simpson in 1755 (printed 1756) first applied the theory to the discussion of errors of observation. The reprint (1757) of this memoir lays down the axioms that positive and negative errors are equally probable, and that certain assignable limits define the range of all errors. Simpson also discusses continuous errors and describes a probability curve.

The first two laws of error that were proposed both originated with Pierre-Simon Laplace. The first law was published in 1774 and stated that the frequency of an error could be expressed as an exponential function of the numerical magnitude of the error, disregarding sign. The second law of error was proposed in 1778 by Laplace and stated that the frequency of the error is an exponential function of the square of the error. The second law of error is called the normal distribution or the Gauss law. "It is difficult historically to attribute that law to Gauss, who in spite of his well-known precocity had probably not made this discovery before he was two years old."

Daniel Bernoulli (1778) introduced the principle of the maximum product of the probabilities of a system of concurrent errors.

Adrien-Marie Legendre (1805) developed the method of least squares, and introduced it in his *Nouvellesméthodes pour la détermination des orbites des comètes* (*New Methods for Determining the Orbits of Comets*). In ignorance of Legendre's contribution, an Irish-American writer, Robert Adrain, editor of "The Analyst" (1808), first deduced the law of facility of error,

$$\phi(x) = ce^{-h^2x^2},$$

where h is a constant depending on precision of observation, and c is a scale factor ensuring that the area under the curve equals 1. He gave two proofs, the second being essentially the same as John Herschel's (1850). Gauss gave the first proof that seems to have been known in Europe (the third after Adrain's) in 1809. Further proofs were given by Laplace (1810, 1812), Gauss (1823), James Ivory (1825, 1826), Hagen (1837), Friedrich Bessel (1838), W. F. Donkin (1844, 1856), and Morgan Crofton (1870). Other contributors were Ellis (1844), De Morgan (1864), Glaisher (1872), and Giovanni Schiaparelli (1875). Peters's (1856) formula for r, the probable error of a single observation, is well known.

In the nineteenth century authors on the general theory included Laplace, SylvestreLacroix (1816), Littrow (1833), AdolpheQuetelet (1853), Richard Dedekind (1860), Helmert (1872), Hermann Laurent (1873), Liagre, Didion, and Karl Pearson. Augustus De Morgan and George Boole improved the exposition of the theory.

Andrey Marko vintroduced the notion of Markov chains (1906), which played an important role in stochastic processes theory and its applications. The modern theory of probability based on the measure theory was developed by Andrey Kolmogorov (1931). On the geometric side contributors to *The Educational Times* were influential (Miller, Crofton, McColl, Wolstenholme, Watson, and Artemas Martin).

Theory

Like other theories, the theory of probability is a representation of probabilistic concepts in formal terms—that is, in terms that can be considered separately from their meaning. These formal terms are manipulated by the rules of mathematics and logic, and any results are interpreted or translated back into the problem domain.

There have been at least two successful attempts to formalize probability, namely the Kolmogorov formulation and the Cox formulation. In Kolmogorov's formulation, sets are interpreted as events and probability itself as a measure on a class of sets. In Cox's theorem, probability is taken as a primitive (that is, not further analyzed) and the emphasis is on constructing a consistent assignment of probability values to propositions. In both cases, the laws of probability are the same, except for technical details. There are other methods for quantifying uncertainty, such as the Dempster-Shafer theory or possibility theory, but those are essentially different and not compatible with the laws of probability as usually understood.

Applications

Probability theory is applied in everyday life in risk assessment and in trade on financial markets. Governments apply probabilistic methods in environmental regulation, where it is called pathway analysis. A good example is the effect of the perceived probability of any widespread Middle East conflict on oil prices—which have ripple effects in the economy as a whole. An assessment by a commodity trader that a war is more likely vs. less likely sends prices up or down, and signals other traders of that opinion. Accordingly, the probabilities are neither assessed independently nor necessarily very rationally. The theory of behavioural finance emerged to describe the effect of such groupthink on pricing, on policy, and on peace and conflict. The discovery of rigorous methods to assess and combine probability assessments has changed society. It is important for most citizens to understand how probability assessments are made, and how they contribute to decisions.

Another significant application of probability theory in everyday life is reliability. Many consumer products, such as automobiles and consumer electronics, use reliability theory in product design to reduce the probability of failure. Failure probability may influence a manufacture's decisions on a product's warranty. The cache language model and other statistical language models that are used in natural language processing are also examples of applications of probability theory.

Mathematical Treatment

Consider an experiment that can produce a number of results. The collection of all results is called the sample space of the experiment. The power set of the sample space is formed by considering all different collections of possible results. For example, rolling a die can produce six possible results. One collection of possible results gives an odd number on the die. Thus, the subset {1,3,5} is an element of the power set of the sample space of die rolls. These collections are called "events." In this case, {1,3,5} is the event that the die falls on some odd number. If the results that actually occur fall in a given event, the event is said to have occurred.

A probability is a way of assigning every event a value between zero and one, with the requirement that the event made up of all possible results (in our example, the event {1,2,3,4,5,6}) is assigned a value of one. To qualify as a probability, the assignment of values must satisfy the requirement that if you look at a collection of mutually exclusive events (events with no common results, e.g., the events {1,6}, {3}, and {2,4} are all mutually exclusive), the probability that at least one of the events will occur is given by the sum of the probabilities of all the individual events.

The probability of an event A is written as P(A), p(A) or Pr(A). This mathematical definition of probability can extend to infinite sample spaces, and even uncountable sample spaces, using the concept of a measure. The *opposite* or *complement* of an event A is the event [not A] (that is, the event of A not occurring); its probability is given by P(not$A$) = 1 - P(A). As an example, the chance of not rolling a six on a six-sided die is 1 – (chance of rolling a six) $=1-\frac{1}{6}=\frac{5}{6}$. If both events A and B occur on a single performance of an experiment, this is called the intersection or joint probability of A and B, denoted as $P(A \cap B)$.

Independent Probability

If two events, A and B are independent then the joint probability is

$$P(A \text{ and } B) = P(A \cap B) = P(A)P(B),$$

for example, if two coins are flipped the chance of both being heads is $\frac{1}{2}\times\frac{1}{2}=\frac{1}{4}$.

Mutually Exclusive

If either event A or event B or both events occur on a single performance of an experiment this is called the union of the events A and B denoted as $P(A \cup B)$. If two events are mutually exclusive then the probability of either occurring is

$$P(A \text{ or } B) = P(A \cup B) = P(A) + P(B).$$

For example, the chance of rolling a 1 or 2 on a six-sided die is $P(1 \text{ or } 2) = P(1) + P(2) = \frac{1}{6} + \frac{1}{6} = \frac{1}{3}$.

Not Mutually Exclusive

If the events are not mutually exclusive then

$$P(A \text{ } or \text{ } B) = P(A) + P(B) - P(A \text{ } and \text{ } B).$$

For example, when drawing a single card at random from a regular deck of cards, the chance of getting a heart or a face card (J,Q,K) (or one that is both) is $\frac{13}{52}+\frac{12}{52}-\frac{3}{52}=\frac{11}{26}$, because of the 52 cards of a deck 13 are hearts, 12 are face cards, and 3 are both: here the possibilities included in the "3 that are both" are included in each of the "13 hearts" and the "12 face cards" but should only be counted once.

Conditional Probability

Conditional probability is the probability of some event A, given the occurrence of some other event B. Conditional probability is written $P(A \quad B)$, and is read "the probability of A, given B". It is defined by

$$P(A \quad B) = \frac{P(A \cap B)}{P(B)}.$$

If $P(B) = 0$ then $P(A \quad B)$ is formally undefined by this expression. However, it is possible to define a conditional probability for some zero-probability events using a σ-algebra of such events (such as those arising from a continuous random variable).

For example, in a bag of 2 red balls and 2 blue balls (4 balls in total), the probability of taking a red ball is $1/2$; however, when taking a second ball, the probability of it being either a red ball or a blue ball depends on the ball previously taken, such as, if a red ball was taken, the probability of picking a red ball again would be $1/3$ since only 1 red and 2 blue balls would have been remaining.

Relation to Eandomness

In a deterministic universe, based on Newtonian concepts, there would be no probability if all conditions are known (Laplace's demon), (but there are situations in which sensitivity to initial conditions exceeds our ability to measure them, i.e. know them). In the case of a roulette wheel, if the force of the hand and the period of that force are known, the number on which the ball will stop would be a certainty (though as a practical matter, this would likely be true only of a roulette wheel that had not been exactly levelled — as Thomas A. Bass' Newtonian Casino revealed). Of course, this also assumes knowledge of inertia and friction of the wheel, weight, smoothness and roundness of the ball, variations in hand speed during the turning and so forth. A probabilistic description can thus be more useful than Newtonian mechanics for analyzing the pattern of outcomes of repeated rolls of roulette wheel. Physicists face the same situation in kinetic theory of gases, where the system, while deterministic *in principle*, is so complex (with the number of molecules typically the order of magnitude of Avogadro constant $6.02 \cdot 10^{23}$) that only statistical description of its properties is feasible.

Probability theory is required to describe quantum phenomena. A revolutionary discovery of early 20th century physics was the random character of all physical processes that occur at sub-atomic scales and are governed by the laws of quantum mechanics. The objective wave function evolves deterministically but, according to the Copenhagen interpretation, it deals with probabilities of observing, the outcome being explained by a wave function collapse when an observation is made. However, the loss of determinism for the sake of instrumentalism did not meet with universal approval. Albert Einstein famously remarked in a letter to Max Born: "I am convinced that God does not play dice". Like Einstein, Erwin Schrödinger, who discovered the wave function, believed quantum mechanics is a statistical approximation of an underlying deterministic reality. In modern interpretations, quantum decoherence accounts for subjectively probabilistic behaviour.

Sample Space

In probability theory, the sample space or universal sample space, often denoted S, Ω, or U (for "universe"), of an experiment or random trial is the set of all possible outcomes. For example, if the experiment is tossing a coin, the sample space is the set {head, tail}. For tossing two coins, the sample space is {(head,head), (head,tail), (tail,head), (tail,tail)}. For tossing a single six-sided die, the sample space is {1, 2, 3, 4, 5, 6}. For some kinds of experiments, there may be two or more plausible sample spaces available. For example, when drawing a card from a standard deck of 52 playing cards, one possibility for the sample space could be the rank (Ace through King), while another could be the suit (clubs, diamonds, hearts, or spades). A complete description of outcomes, however, would specify both the denomination and the suit, and a sample space describing each individual card can be constructed as the Cartesian product of the two sample spaces noted above.

In an elementary approach to probability, any subset of the sample space is usually called an event. However, this gives rise to problems when the sample space is infinite, so that a more precise definition of event is necessary. Under this definition only measurable subsets of the sample space, constituting as-algebra over the sample space itself, are considered events. However, this has essentially only theoretical significance, since in general the σ-algebra can always be defined to include all subsets of interest in applications.

Chapter

9

Methods of Matrix Inversion

Gaussian Elimination

Gauss–Jordan elimination is an algorithm that can be used to determine whether a given matrix is invertible and to find the inverse. An alternative is the LU decomposition which generates upper and lower triangular matrices which are easier to invert. For special purposes, it may be convenient to invert matrices by treating $m \cdot n$-by-$m \cdot n$ matrices as m-by-m matrices of n-by-n matrices, and applying one or another formula recursively (other sized matrices can be padded out with dummy rows and columns).

For other purposes, a variant of Newton's method may be convenient. Newton's method is particularly useful when dealing with families of related matrices: sometimes a good starting point for refining an approximation for the new inverse can be the already obtained inverse of a previous matrix that nearly matches the current matrix. Newton's method is also useful for "touch up" corrections to the Gauss–Jordan algorithm which has been contaminated by small errors due to imperfect computer arithmetic.

Eigende composition

If matrix A can be eigende composed and if none of its eigenvalues are zero, then A is nonsingular and its inverse is given by

$$A^{-1} = Q\Lambda^{-1}Q^{-1}$$

Furthermore, because Λ is a diagonal matrix, its inverse is easy to calculate:

$$\left[\Lambda^{-1}\right]_{ii} = \frac{1}{\lambda_i}$$

Cholesky Decomposition

If matrix A is positive definite, then its inverse can be obtained as

$$A^{-1} = (L^*)^{-1} L^{-1},$$

where L is the lower triangular Cholesky decomposition of A.

Analytic Solution

Writing the transpose of the matrix of cofactors, known as an adjugate matrix, can also be an efficient way to calculate the inverse of *small* matrices, but this recursive method is inefficient for large matrices. To determine the inverse, we calculate a matrix of cofactors:

$$A^{-1} = \frac{1}{|A|}\left(C^T\right)_{ij} = \frac{1}{|A|}\left(C_{ji}\right) = \frac{1}{|A|}\begin{pmatrix} C_{11} & C_{21} & \cdots & C_{n1} \\ C_{12} & C_{22} & \cdots & C_{n2} \\ \vdots & \vdots & \ddots & \vdots \\ C_{1n} & C_{2n} & \cdots & C_{nn} \end{pmatrix}$$

where $|A|$ is the determinant of A, C_{ij} is the matrix of cofactors, and C^T represents the matrix transpose.

Inversion of 2×2 Matrices

The *cofactor equation* listed above yields the following result for 2×2 matrices. Inversion of these matrices can be done easily as follows:

$$A^{-1} = \begin{bmatrix} a & b \\ c & d \end{bmatrix}^{-1} = \frac{1}{\det(A)}\begin{bmatrix} d & -b \\ -c & a \end{bmatrix} = \frac{1}{ad-bc}\begin{bmatrix} d & -b \\ -c & a \end{bmatrix}.$$

This is possible because $1/(ad-bc)$ is the reciprocal of the determinant of the matrix in question, and the same strategy could be used for other matrix sizes.

Inversion of 3×3 Matrices

A computationally efficient 3×3 matrix inversion is given by

$$A^{-1} = \begin{bmatrix} a & b & c \\ d & e & f \\ g & h & k \end{bmatrix}^{-1} = \frac{1}{\det(A)}\begin{bmatrix} A & B & C \\ D & E & F \\ G & H & K \end{bmatrix}^T = \frac{1}{\det(A)}\begin{bmatrix} A & D & G \\ B & E & H \\ C & F & K \end{bmatrix}$$

where the determinant of A can be computed by applying the rule of Sarrus as follows:

$$\det(A) = a(ek - fh) - b(kd - fg) + c(dh - eg).$$

If the determinant is non-zero, the matrix is invertible, with the elements of the above matrix on the right side given by

$$\begin{array}{lll} A = (ek - fh) & D = -(bk - ch) & G = (bf - ce) \\ B = -(dk - fg) & E = (ak - cg) & H = -(af - cd) \\ C = (dh - eg) & F = -(ah - bg) & K = (ae - bd) \end{array}$$

The general 3×3 inverse can be expressed concisely in terms of the cross product and triple product:

If a matrix $A = [x_0, x_1, x_2]$ (consisting of three column vectors, x_0, x_1, and x_2) is invertible, its inverse is given by

$$A^{-1} = \frac{1}{\det(A)} \begin{bmatrix} (x_1 \times x_2)^T \\ (x_2 \times x_0)^T \\ (x_0 \times x_1)^T \end{bmatrix}.$$

Note that $\det(A)$ is equal to the triple product of x_0, x_1, and x_2—the volume of the parallelepiped formed by the rows or columns:

$$\det(A) = x_0 \cdot (x_1 \times x_2).$$

The correctness of the formula can be checked by using cross- and triple-product properties and by noting that for groups, left and right inverses always coincide. Intuitively, because of the cross products, each row of A^{-1} is orthogonal to the non-corresponding two columns of A (causing the off-diagonal terms of $I = A^{-1}A$ be zero). Dividing by

$$\det(A) = x_0 \cdot (x_1 \times x_2)$$

causes the diagonal elements of $I = A^{-1}A$ to be unity. For example, the first diagonal is:

$$1 = \frac{1}{x_0 \cdot (x_1 \times x_2)} x_0 \cdot (x_1 \times x_2).$$

Blockwise Inversion

Matrices can also be *inverted blockwise* by using the following analytic inversion formula:

$$\begin{bmatrix} A & B \\ C & D \end{bmatrix}^{-1} = \begin{bmatrix} A^{-1} + A^{-1}B(D - CA^{-1}B)^{-1}CA^{-1} & -A^{-1}B(D - CA^{-1}B)^{-1} \\ -(D - CA^{-1}B)^{-1}CA^{-1} & (D - CA^{-1}B)^{-1} \end{bmatrix}$$

where A, B, C and D are matrix sub-blocks of arbitrary size. (A and D must be square, so that they can be inverted. Furthermore, A and $D–CA^{-1}B$ must be nonsingular.)

This strategy is particularly advantageous if A is diagonal and $D–CA^{-1}B$ (the Schur complement of A) is a small matrix, since they are the only matrices requiring inversion. This technique was reinvented several times and is due to Hans Boltz (1923), who used it for the inversion of geodetic matrices, and Tadeusz Banachiewicz (1937), who generalized it and proved its correctness. The nullity theorem says that the nullity of A equals the nullity of the sub-block in the lower right of the inverse matrix, and that the nullity of B equals the nullity of the sub-block in the upper right of the inverse matrix.

The inversion procedure that led to Equation (1) performed matrix block operations that operated on C and D first. Instead, if A and B are operated on first, and provided D and $A–BD^{-1}C$ are nonsingular , the result is

$$\begin{bmatrix} A & B \\ C & D \end{bmatrix}^{-1} = \begin{bmatrix} (A-BD^{-1}C)^{-1} & -(A-BD^{-1}C)^{-1}BD^{-1} \\ -D^{-1}C(A-BD^{-1}C)^{-1} & D^{-1}+D^{-1}C(A-BD^{-1}C)^{-1}BD^{-1} \end{bmatrix}.$$

Equating Equations (1) and (2) leads to

$$(A-BD^{-1}C)^{-1} = A^{-1} + A^{-1}B(D-CA^{-1}B)^{-1}CA^{-1}$$

$$(A-BD^{-1}C)^{-1}BD^{-1} = A^{-1}B(D-CA^{-1}B)^{-1}$$

$$D^{-1}C(A-BD^{-1}C)^{-1} = (D-CA^{-1}B)^{-1}CA^{-1}$$

$$D^{-1} + D^{-1}C(A-BD^{-1}C)^{-1}BD^{-1} = (D-CA^{-1}B)^{-1}$$

where Equation (3) is the matrix inversion lemma, which is equivalent to the binomial inverse theorem.

Since a blockwise inversion of an $n\times n$ matrix requires inversion of two half-sized matrices and 6 multiplications between two half-sized matrices, and since matrix multiplication algorithm has a lower bound of $\Omega(n^2 \log n)$ operations, it can be shown that a divide and conquer algorithm that uses blockwise inversion to invert a matrix runs with the same time complexity as the matrix multiplication algorithm that is used internally.

By Neumann Series

If a matrix A has the property that

$$\lim_{n\to\infty}(I-A)^n = 0$$

then A is nonsingular and its inverse may be expressed by a Neumann series:

$$A^{-1} = \sum_{n=0}^{\infty}(I - A)^n.$$

Truncating the sum results in an "approximate" inverse which may be useful as a preconditioner.

More generally, if A is "near" the invertible matrix X in the sense that

$$\lim_{n\to\infty}(I - X^{-1}A)^n = 0 \ or \ \lim_{n\to\infty}(I - AX^{-1})^n = 0$$

then A is nonsingular and its inverse is

$$A^{-1} = \sum_{n=0}^{\infty}\left(X^{-1}(X - A)\right)^n X^{-1}.$$

If it is also the case that A–X has rank 1 then this simplifies to

$$A^{-1} = X^{-1} - \frac{X^{-1}(A - X)X^{-1}}{1 + tr(X^{-1}(A - X))}.$$

Derivative of the Matrix Inverse

Suppose that the invertible matrix A depends on a parameter t. Then the derivative of the inverse of A with respect to t is given by

$$\frac{dA^{-1}}{dt} = -A^{-1}\frac{dA}{dt}A^{-1}.$$

To derive the above expression for the derivative of the inverse of A, one can differentiate the definition of the matrix inverse $\mathrm{A}^{-1}\mathrm{A} = \mathrm{I}$ and then solve for the inverse of A:

$$\frac{dA^{-1}A}{dt} = \frac{dA^{-1}}{dt}A + A^{-1}\frac{dA}{dt} = \frac{dI}{dt} = 0.$$

Subtracting $A^{-1}\frac{dA}{dt}$ from both sides of the above and multiplying on the right by A^{-1} gives the correct expression for the derivative of the inverse:

$$\frac{dA^{-1}}{dt} = -A^{-1}\frac{dA}{dt}A^{-1}.$$

Similarly, if ϵ is a small number then

$$(A + \epsilon X)^{-1} = A^{-1} - \epsilon A^{-1}XA^{-1} + \mathcal{O}(\epsilon^2).$$

Moore–Penrose Pseudoinverse

Some of the properties of inverse matrices are shared by Moore–Penrose pseudoinverses, which can be defined for any m-by-n matrix.

Applications

For most practical applications, it is *not* necessary to invert a matrix to solve a system of linear equations; however, for a unique solution, it *is* necessary that the matrix involved be invertible.

Decomposition techniques like LU decomposition are much faster than inversion, and various fast algorithms for special classes of linear systems have also been developed.

Matrix Inverses in Real-time Simulations

Matrix inversion plays a significant role in computer graphics, particularly in 3D graphics rendering and 3D simulations. Examples include screen-to-world ray casting, world-to-subspace-to-world object transformations, and physical simulations.

Matrix Inverses in MIMO Wireless Communication

Matrix inversion also play a significant role in the MIMO (Multiple-Input, Multiple–Output) technology in wireless communications. The MIMO system consists of N transmit and M receive antennas. Unique signals, occupying the same frequency band, are sent via N transmit antennas and are received via M receive antennas. The signal arriving at each receive antenna will be a linear combination of the N transmitted signals forming a N×M transmission matrix H. It is crucial for the matrix H to be invertible for the receiver to be able to figure out the transmitted information.

Linear Programming

Linear programming (LP, or linear optimization) is a mathematical method for determining a way to achieve the best outcome (such as maximum profit or lowest cost) in a given mathematical model for some list of requirements represented as linear relationships. Linear programming is a specific case of mathematical programming (mathematical optimization).

More formally, linear programming is a technique for the optimization of a linear objective function, subject to linear equality and linear inequality constraints. Its feasible region is a convex polyhedron, which is a set defined as the intersection of finitely many half spaces, each of which is defined by a linear inequality. Its objective function is a real-valued affine function defined on this polyhedron. A linear programming algorithm finds a point in the polyhedron where this function has the smallest (or largest) value if such a point exists.

Linear programmes are problems that can be expressed in canonical form:

$$\begin{array}{ll} \max\mathit{imize} & c^T x \\ \mathit{subject\ to} & Ax \le b \\ \mathit{and} & x \ge 0 \end{array}$$

where x represents the vector of variables (to be determined), c and b are vectors of (known) coefficients, A is a (known) matrix of coefficients, and $(\cdot)^T$ is the matrix transpose. The expression to be maximized or minimized is called the *objective function* ($c^T x$ in this case) the inequalities $\mathrm{A\ x} \le b$ are the constraints which specify a convex polytope over which the objective function is to be optimized. In this context, two vectors are comparable when they have the same dimensions. If every entry in the first is less-than or equal-to the corresponding entry in the second then we can say the first vector is less-than or equal-to the second vector.

Linear programming can be applied to various fields of study. It is used in business and economics, but can also be utilized for some engineering problems. Industries that use linear programming models include transportation, energy, telecommunications, and manufacturing. It has proved useful in modelling diverse types of problems in planning, routing, scheduling, assignment, and design.

History

The problem of solving a system of linear inequalities dates back at least as far as Fourier, after whom the method of Fourier-Motzkin elimination is named. The earliest linear programming was first developed by Leonid Kantorovich in 1939. Leonid Kantorovich developed the earliest linear programming problems in 1939 for use during World War II to plan expenditures and returns in order to reduce costs to the army and increase losses to the enemy. The method was kept secret until 1947 when George B. Dantzig published the simplex method and John von Neumann developed the theory of duality as a linear optimization solution, and applied it in the field of game theory. Postwar, many industries found its use in their daily planning.

The linear-programming problem was first shown to be solvable in polynomial time by Leonid Khachiyan in 1979, but a larger theoretical and practical breakthrough in the field came in 1984 when Narendra Karmarkar introduced a new interior-point method for solving linear-programming problems.

Dantzig's original example was to find the best assignment of 70 people to 70 jobs. The computing power required to test all the permutations to select the best assignment is vast; the number of possible configurations exceeds the number of particles in the observable universe. However, it takes only a moment to find the optimum solution by posing the problem as a linear programme and applying the simplex algorithm. The theory behind linear programming drastically reduces the number of possible optimal solutions that must be checked.

Uses

Linear programming is a considerable field of optimization for several reasons. Many practical problems in operations research can be expressed as linear programming problems. Certain special cases of linear programming, such as *network flow* problems and *multicommodity flow* problems are considered important enough to have generated much research on specialized algorithms for their solution.

A number of algorithms for other types of optimization problems work by solving LP problems as sub-problems. Historically, ideas from linear programming have inspired many of the central concepts of optimization theory, such as *duality, decomposition,* and the importance of *convexity* and its generalizations. Likewise, linear programming is heavily used in microeconomics and company management, such as planning, production, transportation, technology and other issues. Although the modern management issues are ever-changing, most companies would like to maximize profits or minimize costs with limited resources. Therefore, many issues can be characterized as linear programming problems.

Standard Form

Standard form is the usual and most intuitive form of describing a linear programming problem. It consists of the following three parts:

- A linear function to be maximized

  e.g. $f(x_1,x_2) = c_1x_1 + c_2x_2$

- Problem constraints of the following form

  e.g.

  $$a_{11}x_1 + a_{12}x_2 \le b_1$$
  $$a_{21}x_1 + a_{22}x_2 \le b_2$$
  $$a_{31}x_1 + a_{32}x_2 \le b_3$$

- Non-negative variables

  e.g.

  $$x_1 \ge 0$$
  $$x_2 \ge 0$$

The problem is usually expressed in *matrix form*, and then becomes:

$$\max\{c^T x \mid Ax \le b \wedge x \ge 0\}$$

Other forms, such as minimization problems, problems with constraints on alternative forms, as well as problems involving negative variables can always be rewritten into an equivalent problem in standard form.

Augmented form (Slack Form)

Linear programming problems must be converted into *augmented form* before being solved by the simplex algorithm. This form introduces non-negative *slack variables* to replace inequalities with equalities in the constraints. The problem can then be written in the following block matrix form:

Maximize Z:

$$\begin{bmatrix} 1 & -c^T & 0 \\ 0 & A & I \end{bmatrix} \begin{bmatrix} Z \\ x \\ x_s \end{bmatrix} = \begin{bmatrix} 0 \\ b \end{bmatrix}$$

$$x,\ x_s \ge 0$$

where x_s are the newly introduced slack variables, and Z is the variable to be maximized.

Duality

Every linear programming problem, referred to as a *primal* problem, can be converted into a dual problem, which provides an upper bound to the optimal value of the primal problem. In matrix form, we can express the *primal* problem as:

Maximize c^Tx subject to $Ax \le b$, $x \ge 0$;

with the corresponding symmetric dual problem,

Minimize $b^T y$ subject to $A^T y \geq c$, $y \geq 0$.

An alternative primal formulation is:

Maximize $c^T x$ subject to $Ax \leq b$;

with the corresponding asymmetric dual problem,

Minimize $b^T y$ subject to $A^T y = c$, $y \geq 0$.

There are two ideas fundamental to duality theory. One is the fact that (for the symmetric dual) the dual of a dual linear programme is the original primal linear programme. Additionally, every feasible solution for a linear programme gives a bound on the optimal value of the objective function of its dual. The weak duality theorem states that the objective function value of the dual at any feasible solution is always greater than or equal to the objective function value of the primal at any feasible solution. The strong duality theorem states that if the primal has an optimal solution, x^*, then the dual also has an optimal solution, y^*, such that $c^T x^* = b^T y^*$.

A linear programme can also be unbounded or infeasible. Duality theory tells us that if the primal is unbounded then the dual is infeasible by the weak duality theorem. Likewise, if the dual is unbounded, then the primal must be infeasible. However, it is possible for both the dual and the primal to be infeasible.

Covering-packing Dualities

A covering LP is a linear programme of the form:

Minimize: $b^T y$,

Subject to: $A^T y \geq c$, $y \geq 0$,

such that the matrix A and the vectors b and c are non-negative.

The dual of a covering LP is a packing LP, a linear programme of the form:

Maximize: $c^T x$,

Subject to: $Ax \leq b$, $x \geq 0$,

such that the matrix A and the vectors b and c are non-negative.

Examples

Covering and packing LPs commonly arise as a linear programming relaxation of a combinatorial problem and are important in the study of approximation algorithms. For example, the LP relaxations of the set packing problem, the independent set problem, and the matching problem are packing LPs. The LP relaxations of the set cover problem, the vertex cover problem, and the dominating set problem are also covering LPs.

Finding a fractional colouring of a graph is another example of a covering LP. In this case, there is one constraint for each vertex of the graph and one variable for each independent set of the graph.

Complementary Slackness

It is possible to obtain an optimal solution to the dual when only an optimal solution to the primal is known using the complementary slackness theorem. The theorem states:

Suppose that $x = (x_1, x_2, \dots, x_n)$ is primal feasible and that $y = (y_1, y_2, \dots, y_m)$ is dual feasible. Let $(w_1, w_2, \dots, w_m)$ denote the corresponding primal slack variables, and let $(z_1, z_2, \dots, z_n)$ denote the corresponding dual slack variables. Then x and y are optimal for their respective problems if and only if

- $x_j z_j = 0$, for $j = 1, 2, \dots, n$, and
- $w_i y_i = 0$, for $i = 1, 2, \dots, m$.

So if the i-th slack variable of the primal is not zero, then the i-th variable of the dual is equal to zero. Likewise, if the j-th slack variable of the dual is not zero, then the j-th variable of the primal is equal to zero.

This necessary condition for optimality conveys a fairly simple economic principle. In standard form (when maximizing), if there is slack in a constrained primal resource (i.e., there are "leftovers"), then additional quantities of that resource must have no value. Likewise, if there is slack in the dual (shadow) price non-negativity constraint requirement, i.e., the price is not zero, then there must be scarce supplies (no "leftovers").

Theory

Existence of Optimal Solutions

Geometrically, the linear constraints define the feasible region, which is a convexpolyhedron. A linear function is a convex function, which implies that every local minimum is a global minimum; similarly, a linear function is a concave function, which implies that every local maximum is a global maximum.

Optimal solution need not exist, for two reasons. First, if two constraints are inconsistent, then no feasible solution exists: For instance, the constraints $x \geq 2$ and $x \leq 1$ cannot be satisfied jointly;

in this case, we say that the LP is *infeasible*. Second, when the polytope is unbounded in the direction of the gradient of the objective function (where the gradient of the objective function is the vector of the coefficients of the objective function), then no optimal value is attained.

Optimal Vertices (and Rays) of Polyhedra

Otherwise, if a feasible solution exists and if the (linear) objective function is bounded, then the optimum value is always attained on the boundary of optimal level-set, by the *maximum principle* for *convex functions* (alternatively, by the *minimum* principle for *concave functions*): Recall that linear functions are both convex and concave. However, some problems have distinct optimal solutions: For example, the problem of finding a feasible solution to a system of linear inequalities is a linear programming problem in which the objective function is the zero function (that is, the constant function taking the value zero everywhere): For this feasibility problem with the zero-function for its objective-function, if there are two distinct solutions, then every convex combination of the solutions is a solution.

The vertices of the polytope are also called *basic feasible solutions*. The reason for this choice of name is as follows. Let d denote the number of variables. Then the fundamental theorem of linear inequalities implies (for feasible problems) that for every vertex x^* of the LP feasible region, there exists a set of d (or fewer) inequality constraints from the LP such that, when we treat those d constraints as equalities, the unique solution is x^*. Thereby we can study these vertices by means of looking at certain subsets of the set of all constraints (a discrete set), rather than the continuum of LP solutions. This principle underlies the simplex algorithm for solving linear programmes.

Algorithms

Basis Exchange Algorithms

Simplex Algorithm of Dantzig: The simplex algorithm, developed by George Dantzig in 1947, solves LP problems by constructing a feasible solution at a vertex of the polytope and then walking along a path on the edges of the polytope to vertices with non-decreasing values of the objective function until an optimum is reached for sure. In many practical problems, "stalling" occurs: Many pivots are made with no increase in the objective function. In rare practical problems, the usual versions of the simplex algorithm may actually "cycle". To avoid cycles, researchers developed new pivoting rules.

In practice, the simplex algorithm is quite efficient and can be guaranteed to find the global optimum if certain precautions against *cycling* are taken. The simplex algorithm has been proved to solve "random" problems efficiently, i.e. in a cubic number of steps, which is similar to its behaviour on practical problems.

However, the simplex algorithm has poor worst-case behaviour: Klee and Minty constructed a family of linear programming problems for which the simplex method takes a number of steps exponential in the problem size. In fact, for some time it was not known whether the linear programming problem was solvable in polynomial time, i.e. of complexity class P.

Criss-cross Algorithm

Like the simplex algorithm of Dantzig, the criss-cross algorithm is a basis-exchange algorithm that pivots between bases. However, the criss-cross algorithm need not maintain feasibility, but can pivot rather from a feasible basis to an infeasible basis. The criss-cross algorithm does not have polynomial time-complexity for linear programming. Both algorithms visit all 2^D corners of a (perturbed) cube in dimension D, the Klee–Minty cube, in the worst case.

Conic Sampling Algorithm of Serang

Like other basis-exchange algorithms, Serang's conic sampling algorithm moves between vertices; but where the simplex algorithm moves along edges by removing and adding one basis at a time, the conic sampling method exchanges multiple bases at a time, and is not restricted to moving along edges of the polytope. Starting at a current vertex, the conic sampling method chooses a random vector that improves the objective value without violating any adjacent constraints. The algorithm then travels along this vector until a limiting constraint is encountered.

From this point, the algorithm projects the objective vector orthogonal to this limiting constraint, and moves along this orthogonal projection until a new constraint is reached. This advancement and projection is repeated until a vertex is reached. Then, a new random vector is chosen. This process is repeated until no vector exists that can improve the objective without violating any local constraints, implying optimality. Essentially, the conic sampling method can be thought of as a vertex sampling method that randomly samples from the collection of vertices with improved objective value. If the vertices with superior objective value are sampled in a roughly uniform manner, then the expected runtime is logarithmic in the number of vertices (and thus polynomial). Sampling the vertices in this manner can permit large, beneficial jumps through the interior, and yield a substantial runtime improvent over the simplex method, especially when the number of constraints, and thus the number of potential vertices, is large; however, the tightest existing upper bound on the worst-case complexity of the conic sampling method is still exponential.

Interior Point

Ellipsoid Algorithm, following Khachiyan

This is the first worst-case polynomial-time algorithm for linear programming. To solve a problem which has n variables and can be encoded in L input bits, this algorithm uses $O(n^4L)$ pseudo-arithmetic operations on numbers with $O(L)$ digits. Khachiyan's algorithm and his long standing issue was resolved by Leonid Khachiyan in 1979 with the introduction of the ellipsoid method. The convergence analysis have (real-number) predecessors, notably the iterative methods developed by Naum Z. Shor and the approximation algorithms by Arkadi Nemirovski and D. Yudin.

Projective Algorithm of Karmarkar

Khachiyan's algorithm was of landmark importance for establishing the polynomial-time solvability of linear programmes. The algorithm was not a computational break-through, as the simplex method is more efficient for all but specially constructed families of linear programmes.

However, Khachiyan's algorithm inspired new lines of research in linear programming. In 1984, N. Karmarkar proposed a projective method for linear programming. Karmarkar's algorithm improved on Khachiyan's worst-case polynomial bound (giving $O(n^{3.5}L)$). Karmarkar claimed that his algorithm was much faster in practical LP than the simplex method, a claim that created great interest in interior-point methods.

Path-following Algorithms

In contrast to the simplex algorithm, which finds an optimal solution by traversing the edges between vertices on a polyhedral set, interior-point methods move through the interior of the feasible region. Since then, many interior-point methods have been proposed and analyzed. Early successful implementations were based on *affine scaling* variants of the method. For both theoretical and practical purposes, barrier function or path-following methods have been the most popular since the 1990s.

Comparison of Interior-point Methods Versus Simplex Algorithms

The current opinion is that the efficiency of good implementations of simplex-based methods and interior point methods are similar for routine applications of linear programming. However, for specific types of LP problems, it may be that one type of solver is better than another (sometimes much better).

LP solvers are in widespread use for optimization of various problems in industry, such as optimization of flow in transportation networks.

Open Problems and Recent Work

There are several open problems in the theory of linear programming, the solution of which would represent fundamental breakthroughs in mathematics and potentially major advances in our ability to solve large-scale linear programmes.

- Does LP admit a strongly polynomial-time algorithm?
- Does LP admit a strongly polynomial algorithm to find a strictly complementary solution?
- Does LP admit a polynomial algorithm in the real number (unit cost) model of computation?

This closely related set of problems has been cited by Stephen Smale as among the 18 greatest unsolved problems of the 21st century. In Smale's words, the third version of the problem "is the main unsolved problem of linear programming theory."

While algorithms exist to solve linear programming in weakly polynomial time, such as the ellipsoid methods and interior-point techniques, no algorithms have yet been found that allow strongly polynomial-time performance in the number of constraints and the number of variables. The development of such algorithms would be of great theoretical interest, and perhaps allow practical gains in solving large LPs as well.

Although the Hirsch conjecture was recently disproved for higher dimensions, it still leaves the following questions open.

- Are there pivot rules which lead to polynomial-time Simplex variants?
- Do all polytopal graphs have polynomially bounded diameter?

These questions relate to the performance analysis and development of Simplex-like methods. The immense efficiency of the Simplex algorithm in practice despite its exponential-time theoretical performance hints that there may be variations of Simplex that run in polynomial or even strongly polynomial time. It would be of great practical and theoretical significance to know whether any such variants exist, particularly as an approach to deciding if LP can be solved in strongly polynomial time.

The Simplex algorithm and its variants fall in the family of edge-following algorithms, so named because they solve linear programming problems by moving from vertex to vertex along edges of a polytope. This means that their theoretical performance is limited by the maximum number of edges between any two vertices on the LP polytope. As a result, we are interested in knowing the maximum graph-theoretical diameter of polytopal graphs. It has been proved that all polytopes have subexponential diameter. The recent disproof of the Hirsch conjecture is the first step to prove whether any polytope has superpolynomial diameter. If any such polytopes exist, then no edge-following variant can run in polynomial time. Questions about polytope diameter are of independent mathematical interest.

Simplex pivot methods preserve primal (or dual) feasibility. On the other hand, criss-cross pivot methods do not preserve (primal or dual) feasibility—they may visit primal feasible, dual feasible or primal-and-dual infeasible bases in any order. Pivot methods of this type have been studied since the 1970s. Essentially, these methods attempt to find the shortest pivot path on the arrangement polytope under the linear programming problem. In contrast to polytopal graphs, graphs of arrangement polytopes are known to have small diameter, allowing the possibility of strongly polynomial-time criss-cross pivot algorithm without resolving questions about the diameter of general polytopes.

Integer Unknowns

If all of the unknown variables are required to be integers, then the problem is called an integer programming (IP) or integer linear programming (ILP) problem. In contrast to linear programming, which can be solved efficiently in the worst case, integer programming problems are in many practical situations (those with bounded variables) NP-hard. 0-1 integer programming or binary integer programming (BIP) is the special case of integer programming where variables are required to be 0 or 1 (rather than arbitrary integers). This problem is also classified as NP-hard, and in fact the decision version was one of Karp's 21 NP-complete problems. If only some of the unknown variables are required to be integers, then the problem is called a mixed integer programming (MIP) problem. These are generally also NP-hard because they are even more general than ILP programmes.

There are however some important subclasses of IP and MIP problems that are efficiently solvable, most notably problems where the constraint matrix is totally unimodular and the right-hand sides of the constraints are integers or - more general - where the system has the total dual integrality (TDI) property.

Advanced algorithms for solving integer linear programmes include:

- cutting-plane method
- branch and bound
- branch and cut
- branch and price
- if the problem has some extra structure, it may be possible to apply delayed column generation.

Such integer-programming algorithms are discussed by Padberg and in Beasley.

Integral Linear Programmes

A linear programme in real variables is said to be integral if it has at least one optimal solution which is integral. Likewise, a polyhedron $P = \{x \;\; Ax \geq 0\}$ is said to be integral if for all bounded feasible objective functions *c*, the linear programme $\{\max cx \;\; x \in P\}$ has an optimum x^* with integer coordinates. As observed by Edmonds and Giles in 1977, one can equivalently say that a polyhedron is integral if for every bounded feasible integral objective function *c*, the optimal *value* of the linear programme $\{\max cx \;\; x \in P\}$ is an integer.

Integral linear programmes are of central importance in the polyhedral aspect of combinatorial optimization since they provide an alternate characterization of a problem. Specifically, for any problem, the convex hull of the solutions is an integral polyhedron; if this polyhedron has a nice/compact description, then we can efficiently find the optimal feasible solution under any linear objective. Conversely, if we can prove that a linear programming relaxation is integral, then it is the desired description of the convex hull of feasible (integral) solutions. Note that terminology is not consistent throughout the literature, so one should be careful to distinguish the following two concepts,

- in an *integer linear programme,* described in the previous section, variables are forcibly constrained to be integers, and this problem is NP-hard in general,
- in an *integral linear programme,* described in this section, variables are not constrained to be integers but rather one has proven somehow that the continuous problem always has an integral optimal value (assuming c is integral), and this optimal value may be found efficiently since all polynomial-size linear programmes can be solved in polynomial time.

One common way of proving that a polyhedron is integral is to show that it is totally unimodular. There are other general methods including the integer decomposition property and total dual integrality. Other specific well-known integral LPs include the matching polytope, lattice polyhedra, submodular flow polyhedra, and the intersection of 2 generalized polymatroids/g-polymatroids — e.g. A bounded integral polyhedron is sometimes called a convex lattice polytope, particularly in two dimensions.

Solvers and Scripting (Programming) Languages

Free Open-source Permissive Licenses:

Name	*License*	*Brief info*
JOptimizer	ASL	java convex optimization library - open source
OpenOpt	BSD	Universal cross-platform numerical optimization framework, see its LP page and other problems involved
OPTI Toolbox	BSD	MATLAB Toolbox for solving linear, nonlinear, continuous and discrete optimization problems. See the OPTI LP Examples page for several examples.

Free Open-source Copyleft (Reciprocal) Licenses:

Name	*License*	*Brief info*
Cassowary constraint solver	LGPL	an incremental constraint solving toolkit that efficiently solves systems of linear equalities and inequalities.
glpk	GPL	GNU Linear Programming Kit, a free LP/MILP solver. Uses GNU MathProg modelling language.
Qoca	GPL	a library for incrementally solving systems of linear equations with various goal functions
CLP	CPL	an LP solver from COIN-OR
R-Project	GPL	a programming language and software environment for statistical computing and graphics

MINTO (Mixed Integer Optimizer, an integer programming solver which uses branch and bound algorithm) has publicly available source code but not open source.

Proprietary:

Name	***Brief info***
AIMMS	
AMPL	A popular modelling language for large-scale linear, mixed integer and nonlinear optimisation with a free student version available.
APMonitor	API to MATLAB and Python. Solve an example Linear Programming (LP) problem through a web-interface.
CPLEX	Popular solver with an API for several programming languages, and also has a modelling language and works with AIMMS, AMPL, GAMS, MPL, OpenOpt, OPL Development Studio, and TOMLAB. Free for academic use.
EXCEL Solver Function	
FinMath	A .NET numerical library containing dense and sparse versions of an primal-dual interior-point solver.
FortMP	
GAMS	
Gurobi	Solver with parallel algorithms for large-scale linear programmes, quadratic programmes and mixed-integer programmes. Free for academic use.
IMSL Numerical Libraries	Collections of math and statistical algorithms available in C/C++, Fortran, Java and C#/.NET. Optimization routines in the IMSL Libraries include unconstrained, linearly and nonlinearly constrained minimizations, and linear programming algorithms.
LiPS	Free optimization package intended for solving linear, integer and goal programming problems.
MATLAB	A general-purpose and matrix-oriented programming-language for numerical computing. Linear programming in MATLAB requires the Optimization Toolbox in addition to the base MATLAB product; available routines include BINTPROG and LINPROG

Contd...

Name	*Brief info*
Mathcad	A WYSIWYG math editor. It has functions for solving both linear and nonlinear optimization problems.
Mathematica	A general-purpose programming-language for mathematics, including symbolic and numerical capabilities.
MOSEK	A solver for large scale optimization with API for several languages (C++,java,.net, Matlab and python).
NAG Numerical Library	A collection of mathematical and statistical routines developed by the Numerical Algorithms Group for multiple programming languages (C, C++, Fortran, Visual Basic, Java and C#) and packages (MATLAB, Excel, R, LabVIEW). The Optimization chapter of the NAG Library includes routines for linear programming problems with both sparse and non-sparse linear constraint matrices, together with routines for the optimization of quadratic, nonlinear, sums of squares of linear or nonlinear functions with nonlinear, bounded or no constraints. The NAG Library has routines for both local and global optimization, and for continuous or integer problems.
NMath Stats	A general-purpose .NET statistical library containing a simplex solver.
Optimj	A Java-based modelling language for optimization with a free version available.
SAS/OR	A suite of solvers for Linear, Integer, Nonlinear, Derivative-Free, Network, Combinatorial and Constraint Optimization; the Algebraic modelling languageOPTMODEL; and a variety of vertical solutions aimed at specific problems/markets, all of which are fully integrated with the SAS System.
SCIP	A general-purpose constraint integer programming solver with an emphasis on MIP. Compatible with Zimpl modelling language. Free for academic use and available in source code.
SuanShu	A Java-based math library that supports linear programming and other kinds of numerical optimization.
UFFLP	An easy API for Mixed, Integer and Linear Programming
VisSim	A visual block diagram language for simulation of dynamical systems.

Set (Mathematics)

This article is about what mathematicians call "intuitive" or "naive" set theory.

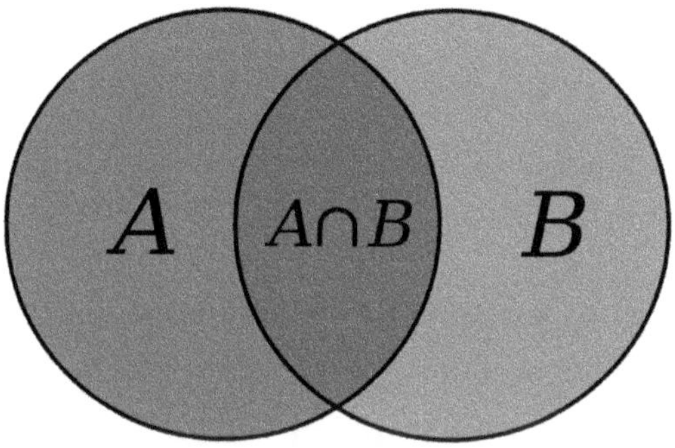

Figure: *An example of a Venn diagram*

The *intersection* of two sets is made up with the objects contained in both sets. In mathematics, a set is a collection of distinct objects, considered as an object in its own right. For example, the numbers 2, 4, and 6 are distinct objects when considered separately, but when they are considered collectively they form a single set of size three, written {2,4,6}. Sets are one of the most fundamental concepts in mathematics. Developed at the end of the 19th century, set theory is now a ubiquitous part of mathematics, and can be used as a foundation from which nearly all of mathematics can be derived. In mathematics education, elementary topics such as Venn diagrams are taught at a young age, while more advanced concepts are taught as part of a university degree.

Definition

A set is a well defined collection of objects. The objects that make up a set (also known as the elements or members of a set) can be anything: numbers, people, letters of the alphabet, other sets, and so on. Georg Cantor, the founder of set theory, gave the following definition of a set at the beginning of his *Beiträge zur Begründung der transfiniten Mengenlehre*:

A set is a gathering together into a whole of definite, distinct objects of our perception [Anschauung] and of our thought – which are called elements of the set.

Sets are conventionally denoted with capital letters. Sets A and B are equal if and only if they have precisely the same elements.

As discussed below, the definition given above turned out to be inadequate for formal mathematics; instead, the notion of a "set" is taken as an undefined primitive in axiomatic set theory, and its properties are defined by the Zermelo–Fraenkel axioms. The most basic properties are that a set "has" elements, and that two sets are equal (one and the same) if and only if every element of one is an element of the other.

Describing Sets

There are two ways of describing, or specifying the members of, a set. One way is by intensional definition, using a rule or semantic description:

A is the set whose members are the first four positive integers.

B is the set of colours of the French flag.

The second way is by extension – that is, listing each member of the set. An extensional definition is denoted by enclosing the list of members in curly brackets:

$C = \{4, 2, 1, 3\}$

$D = \{\text{blue, white, red}\}$.

Every element of a set must be unique; no two members may be identical. (A multiset is a generalized concept of a set that relaxes this criterion.) All set operations preserve this property. The order in which the elements of a set or multiset are listed is irrelevant (unlike for a sequence or tuple). Combining these two ideas into an example

$$\{6, 11\} = \{11, 6\} = \{11, 6, 6, 11\}$$

because the extensional specification means merely that each of the elements listed is a member of the set.

For sets with many elements, the enumeration of members can be abbreviated. For instance, the set of the first thousand positive integers may be specified extensionally as:

$$\{1, 2, 3, \ldots, 1000\},$$

where the ellipsis ("...") indicates that the list continues in the obvious way. Ellipses may also be used where sets have infinitely many members. Thus the set of positive even numbers can be written as $\{2, 4, 6, 8, \ldots\}$.

The notation with braces may also be used in an intensional specification of a set. In this usage, the braces have the meaning "the set of all ...". So, $E = \{\text{playing card suits}\}$ is the set whose four members are ♠, ♦, ♥, and ♣. A more general form of this is set-builder notation, through which, for instance, the set F of the twenty smallest integers that are four less than perfect squares can be denoted:

$$F = \{n^2 - 4 : n \text{ is an integer; and } 0 \leq n \leq 19\}.$$

In this notation, the colon (":") means "such that", and the description can be interpreted as "F is the set of all numbers of the form $n^2 - 4$, such that n is a whole number in the range from 0 to 19 inclusive." Sometimes the vertical bar ("|") is used instead of the colon. One often has the choice of specifying a set intensionally or extensionally. In the examples above, for instance, $A = C$ and $B = D$.

Membership

The key relation between sets is *membership* – when one set is an element of another. If a is a member of B, this is denoted $a \in B$, while if c is not a member of B then $c \notin B$. For example, with respect to the sets $A = \{1,2,3,4\}$, $B = \{\text{blue, white, red}\}$, and $F = \{n^2 - 4 : n$ is an integer; and $0 \leq n \leq 19\}$ defined above,

$4 \in A$ and $12 \in F$; but

$9 \notin F$ and green $\notin B$.

Subsets

If every member of set A is also a member of set B, then A is said to be a *subset* of B, written $A \subseteq B$ (also pronounced *A is contained in B*). Equivalently, we can write $B \supseteq A$, read as *B is a superset of A*, *B includes A*, or *B contains A*. The relationship between sets established by $\subseteq$ is called *inclusion* or *containment*.

If A is a subset of, but not equal to, B, then A is called a *proper subset* of B, written $A \subsetneq B$ (*A is a proper subset of B*) or $B \supsetneq A$ (*B is a proper superset of A*).

Note that the expressions $A \subset B$ and $B \supset A$ are used differently by different authors; some authors use them to mean the same as $A \subseteq B$ (respectively $B \supseteq A$), whereas other use them to mean the same as $A \subsetneq B$ (respectively $B \supsetneq A$).

Example:

- The set of all men is a proper subset of the set of all people.
- $\{1, 3\} \subsetneq \{1, 2, 3, 4\}$.
- $\{1, 2, 3, 4\} \subseteq \{1, 2, 3, 4\}$.

The empty set is a subset of every set and every set is a subset of itself:

- $\varnothing \subseteq A$.
- $A \subseteq A$.

An obvious but useful identity, which can often be used to show that two seemingly different sets are equal:

- $A = B$ if and only if $A \subseteq B$ and $B \subseteq A$.

A partition of a set$S$ is a set of nonempty subsets of S such that every element x in S is in exactly one of these subsets.

Power Sets

The power set of a set S is the set of all subsets of S, including S itself and the empty set. For example, the power set of the set {1, 2, 3} is {{1, 2, 3}, {1, 2}, {1, 3}, {2, 3}, {1}, {2}, {3}, $\varnothing$}. The power set of a set S usually written as $P(S)$.

The power set of a finite set with n elements has 2^n elements. This relationship is one of the reasons for the terminology *power set*. For example, the set {1, 2, 3} contains three elements, and the power set shown above contains $2^3 = 8$ elements.

The power set of an infinite (either countable or uncountable) set is always uncountable. Moreover, the power set of a set is always strictly "bigger" than the original set in the sense that there is no way to pair the elements of a set S with the elements of its power set $P(S)$ such that every element of S set is paired with exactly one element of $P(S)$, and every element of $P(S)$ is paired with exactly one element of S. (There is never a bijection from S onto $P(S)$.)

Every partition of a set S is a subset of the powerset of S.

Cardinality

The cardinality $| S |$ of a set S is "the number of members of S." For example, if B = {*blue, white, red*}, $| B | = 3$.

There is a unique set with no members and zero cardinality, which is called the *empty set* (or the *null set*) and is denoted by the symbol $\varnothing$. For example, the set of all three-sided squares has zero members and thus is the empty set. Though it may seem trivial, the empty set, like the number zero, is important in mathematics; indeed, the existence of this set is one of the fundamental concepts of axiomatic set theory.

Some sets have infinite cardinality. The set N of natural numbers, for instance, is infinite. Some infinite cardinalities are greater than others. For instance, the set of real numbers has greater cardinality than the set of natural numbers. However, it can be shown that the cardinality of (which is to say, the number of points on) a straight line is the same as the cardinality of any segment of that line, of the entire plane, and indeed of any finite-dimensionalEuclidean space.

Special Sets

There are some sets which hold great mathematical importance and are referred to with such regularity that they have acquired special names and notational conventions to identify them. One of these is the empty set, denoted or $\varnothing$. Another is the unit set $\{x\}$ which contains exactly one element, namely x. Many of these sets are represented using blackboard bold or bold typeface. Special sets of numbers include:

- P or $\mathbb{P}$ $\rightarrow$, denoting the set of all primes: $P = \{2, 3, 5, 7, 11, 13, 17, ...\}$.
- N or $\mathbb{N}$, denoting the set of all natural numbers: $N = \{1, 2, 3, . . .\}$ (sometimes defined containing 0).
- Z or $\mathbb{Z}$, denoting the set of all integers (whether positive, negative or zero): $Z = \{..., -2, -1, 0, 1, 2, ...\}$.
- Q or $\mathbb{Q}$, denoting the set of all rational numbers (that is, the set of all proper and improper fractions): $Q = \{a/b : a, b \in Z, b \neq 0\}$. For example, $1/4 \in Q$ and $11/6 \in Q$. All integers are in this set since every integer a can be expressed as the fraction $a/1$ $(Z \subsetneq Q)$.
- R or $\mathbb{R}$, denoting the set of all real numbers. This set includes all rational numbers, together with all irrational numbers (that is, numbers which cannot be rewritten as fractions, such as $\sqrt{2}$, as well as transcendental numbers such as π, e and numbers that cannot be defined).
- C or $\mathbb{C}$, denoting the set of all complex numbers: $C = \{a + bi : a, b \in R\}$. For example, $1 + 2i \in C$.
- H or $\mathbb{H}$, denoting the set of all quaternions: $H = \{a + bi + cj + dk : a, b, c, d \in R\}$. For example, $1 + i + 2j - k \in H$.

Positive and negative sets are denoted by a superscript – or +, for example: $\mathbb{Q}^+$ represents the set of positive rational numbers. Each of the above sets of numbers has an infinite number of elements, and each can be considered to be a proper subset of the sets listed below it. The primes are used less frequently than the others outside of number theory and related fields.

Basic Operations

There are several fundamental operations for constructing new sets from given sets.

Unions

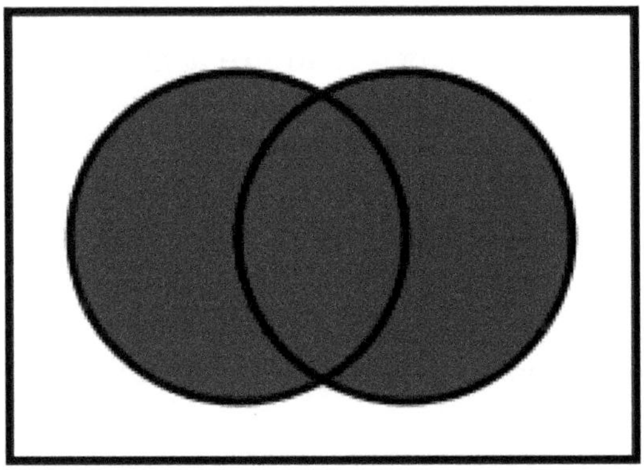

***Figure:** The union of* A *and* B, *denoted* A∪B

Two sets can be "added" together.

The *union* of A and B, denoted by $A \cup B$, is the set of all things which are members of either A or B.

Examples:

- {1, 2} ∪ {red, white} ={1, 2, red, white}.
- {1, 2, green} ∪ {red, white, green} ={1, 2, red, white, green}.
- {1, 2} ∪ {1, 2} = {1, 2}.

Some basic properties of unions:

- $A \cup B = B \cup A$.
- $A \cup (B \cup C) = (A \cup B) \cup C$.
- $A \subseteq (A \cup B)$.
- $A \cup A = A$.
- $A \cup \varnothing = A$.
- $A \subseteq B$if and only if$A \cup B = B$.

Intersections

A new set can also be constructed by determining which members two sets have "in common".

The *intersection* of A and B, denoted by $A \cap B = \varnothing$ is the set of all things which are members of both A and B. If $A \cap B = \varnothing$, then A and B are said to be *disjoint*.

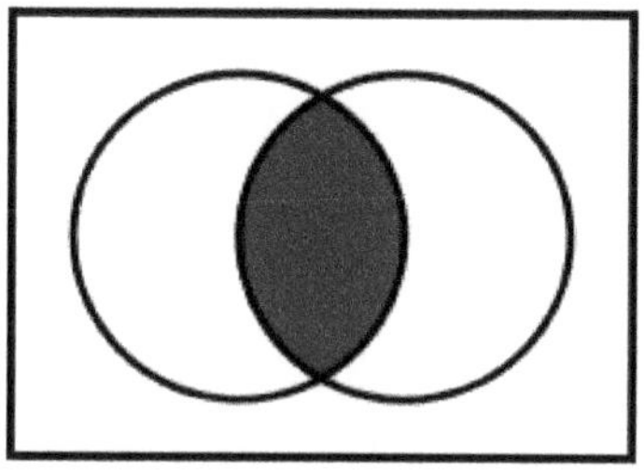

***Figure:** The intersection of A and B, denoted $A \cap B$.*

Examples:

- $\{1, 2\} \cap \{\text{red, white}\} = \varnothing$.
- $\{1, 2, \text{green}\} \cap \{\text{red, white, green}\} = \{\text{green}\}$.
- $\{1, 2\} \cap \{1, 2\} = \{1, 2\}$.

Some basic properties of intersections:

- $A \cap B = B \cap A$.
- $A \cap (B \cap C) = (A \cap B) \cap C$.
- $A \cap B \subseteq A$.
- $A \cap A = A$.
- $A \cap \varnothing = \varnothing$.
- $A \subseteq B$ if and only if $A \cap B = A$.

Complements

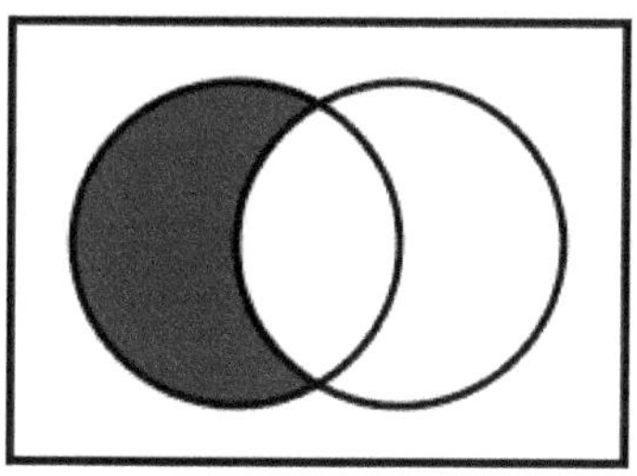

***Figure:** The relative complement*

of B in A

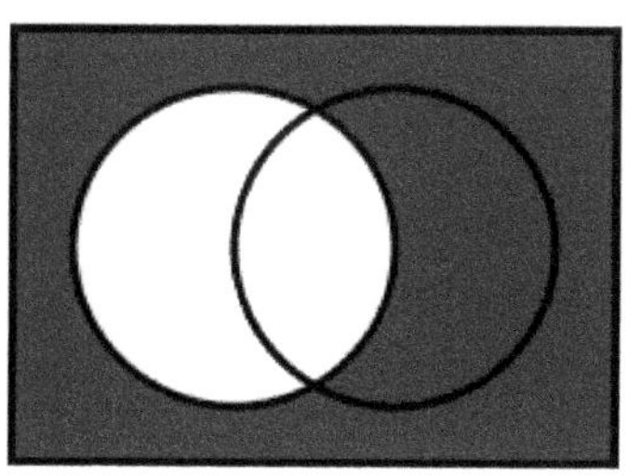

***Figure:** The complement of* A *in* U

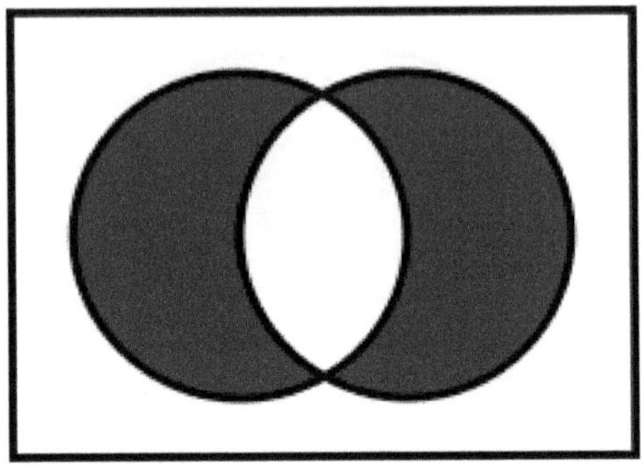

Figure: *The symmetric difference of A and B*

Two sets can also be "subtracted". The *relative complement* of B in A (also called the *set-theoretic difference* of A and B), denoted by $A \setminus B$ (or $A - B$), is the set of all elements which are members of A but not members of B.

Note that it is valid to "subtract" members of a set that are not in the set, such as removing the element *green* from the set {1, 2, 3}; doing so has no effect.

In certain settings all sets under discussion are considered to be subsets of a given universal set U. In such cases, $U \setminus A$ is called the *absolute complement* or simply *complement* of A, and is denoted by $A2$.

Examples:

- $\{1, 2\} \cap \{\text{red, white}\} = \{1, 2\}$.
- $\{1, 2, \text{green}\} \setminus \{\text{red, white, green}\} = \{1, 2\}$.
- $\{1, 2\} \setminus \{1, 2\} = \varnothing$.
- $\{1, 2, 3, 4\} \setminus \{1, 3\} = \{2, 4\}$.
- If U is the set of integers, E is the set of even integers, and O is the set of odd integers, then $U \setminus E = E2 = O$.

Some basic properties of complements:

- $A \setminus B \neq B \setminus A$ for $A \neq B$.
- $A \cup A2 = U$.
- $A \cap A2 = \varnothing$.
- $(A') = A$.
- $A \setminus A = \varnothing$.
- $U2 = \varnothing$ and $\varnothing' = U$.
- $A \setminus B = A \cap B'$.

An extension of the complement is the symmetric difference, defined for sets A, B as

$$A \Delta B = (A \setminus B) \cup (B \setminus A).$$

For example, the symmetric difference of {7,8,9,10} and {9,10,11,12} is the set {7,8,11,12}.

Cartesian Product

A new set can be constructed by associating every element of one set with every element of another set. The *Cartesian product* of two sets A and B, denoted by $A \times B$ is the set of all ordered pairs (a, b) such that a is a member of A and b is a member of B.

Examples:

- {1, 2} × {red, white} = {(1, red), (1, white), (2, red), (2, white)}.
- {1, 2, green} × {red, white, green} = {(1, red), (1, white), (1, green), (2, red), (2, white), (2, green), (green, red), (green, white), (green, green)}.
- {1, 2} × {1, 2} = {(1, 1), (1, 2), (2, 1), (2, 2)}.

Some basic properties of cartesian products:

- $A \times \varnothing = \varnothing$.
- $A \times (B \cup C) = (A \times B) \cup (A \times C)$.
- $(A \cup B) \times C = (A \times C) \cup (B \times C)$.

Let A and B be finite sets. Then

- $|A \times B| = |B \times A| = |A| \times |B|$.

Applications

Set theory is seen as the foundation from which virtually all of mathematics can be derived. For example, structures in abstract algebra, such as groups, fields and rings, are sets closed under one or more operations.

One of the main applications of naive set theory is constructing relations. A relation from a domain A to a codomain B is a subset of the Cartesian product $A \times B$. Given this concept, we are quick to see that the set F of all ordered pairs (x, x^2), where x is real, is quite familiar. It has a domain set R and a codomain set that is also R, because the set of all squares is subset of the set of all reals. If placed in functional notation, this relation becomes $f(x) = x^2$. The reason these two are equivalent is for any given value, y that the function is defined for, its corresponding ordered pair, (y, y^2) is a member of the set F.

Axiomatic Set Theory

Although initially naive set theory, which defines a set merely as *any well-defined* collection, was well accepted, it soon ran into several obstacles. It was found that this definition spawned several paradoxes, most notably:

- Russell's paradox—It shows that the "set of all sets which *do not contain themselves*," i.e. the "set" $\{ x : x \text{ is a set and } x \notin x \}$ does not exist.
- Cantor's paradox—It shows that "the set of all sets" cannot exist.

The reason is that the phrase *well-defined* is not very well defined. It was important to free set theory of these paradoxes because nearly all of mathematics was being redefined in terms of set theory. In an attempt to avoid these paradoxes, set theory was axiomatized based on first-order logic, and thus axiomatic set theory was born.

For most purposes however, naive set theory is still useful.

Principle of Inclusion and Exclusion

This principle gives us the cardinality of the union of sets. |A1 ∪ A2 ∪ A3 ∪ A4 ∪ ... ∪ An|=(|A1| + |A2| + |A3| +...+ |An|)–(|A1 ∩ A2| +|A1 ∩ A3| + ...+|An–1 ∩ An|) + ...+(–1)^{n–1}(|A1 ∩ A2 ∩ A3 ∩.....∩ An|)

Bibliography

Asmussen S.: *Ruin Probabilities*, World Scientific Publishing Co.. Singapore, New York, 2000.

Bernd Gärtner, Jiøí Matoušek: *Understanding and Using Linear Programming*, Springer, Berlin, 2006.

Boyer, Carl B.: *The history of the Calculus and its Conceptual Development*, Dover Publications, New York, 1959.

Brown, William A.: *Matrices and Vector Spaces*, M. Dekker, New York, 1991.

Courant, Richard: *Differential and Integral Calculus. Vol. II*, Wiley Classics Library, New York: John Wiley & Sons, 1988.

Doolittle, R. F.: *Probability and the Origin of Life,* Scientists Confront Creationists, New York, 1983.

Fudenberg, Drew; Tirole, Jean: *Game Theory*, MIT Press, 1983.

Gerber, H.U.: *An Introduction to Mathematical Risk Theory*, Cambridge Univ. Press, Cambridge, 1979.

Golub, Gene H.; Van Loan, Charles F.: *Matrix Computations,* Johns Hopkins, 1996.

Henk Tijms: *Understanding Probability*, Cambridge Univ. Press, Cambridge, 2004.

Hildebrand, F. B.: *Finite-Difference Equations and Simulations*, Prentice-Hall, Englewood Cliffs, New Jersey, 1968.

Horn, Roger A.; Johnson, Charles R.: *Matrix Analysis*, Cambridge University Press, 1985.

Karl-Heinz Borgwardt: *The Simplex Algorithm: A Probabilistic Analysis*, Springer-Verlag, NY, 1987.

Lang, Serge: *Analysis II*, Addison-Wesley, USA, 1969.

Larson, Ron; Hostetler, Robert P.; and Edwards, Bruce H.: *Calculus of a Single Variable: Early Transcendental Functions*, Houghton Mifflin Company, 2003.

Mark de Berg, Marc van Kreveld, Mark Overmars, and Otfried Schwarzkopf: *Computational Geometry,* Springer-Verlag, NY, 2000.

Michael J. Crowe: *A History of Vector Analysis: The Evolution of the Idea of a Vectorial System*, Dover Publications; New York, 1967.

Michel Loeve: *Probability Theory,* Springer-Verlag, New York, 1977.

Navidi, William: *Statistics for Engineers and Scientists*, McGraw-Hill, New York, 2006.

Popper, Karl.: *The Propensity Interpretation of the Calculus of Probability and of the Quantum Theory*, Buttersworth Scientific Publications, 1957.

Press, William H.; Flannery, Brian P.; Teukolsky, Saul A.; Vetterling, William T.: "LU Decomposition and Its Applications", *Numerical Recipes in FORTRAN: The Art of Scientific Computing*, Cambridge University Press, 1992.

Richtmeyer, D. and Morton, K.W.: *Difference Methods for Initial Value Problems*, Wiley, New York, 1967.

Schey, H. M.: *Div Grad Curl and all that: An Informal Text on Vector Calculus,* W. W. Norton & Company, UK, 2005.

Solin, Pavel: *Partial Differential Equations and the Finite Element Method*, Wiley-Interscience, 2005.

Stinson, Douglas R.: *Cryptography*, Discrete Mathematics and its Applications, Chapman & Hall/CRC, 2005.

Stoer, Josef; Bulirsch, Roland: *Introduction to Numerical Analysis,* Springer-Verlag, DE; New York, 2002.

Translation: Kolmogorov: *Foundations of the Theory of Probability*, Chelsea, New York, 1956.

Vazirani, Vijay V.: *Approximation Algorithms*, Springer-Verlag, NY, 2001.

Wang, Zhenyuan, and, George J. Klir.: *Fuzzy Measure Theory*, Plenum Press, New York, 1991.

Index

I

L

M

N

O

P

Q

R

S

T

U

V

W

❑❑❑

www.ingramcontent.com/pod-product-compliance
Ingram Content Group UK Ltd.
Pitfield, Milton Keynes, MK11 3LW, UK
UKHW021328190726
13845UKWH00004B/77

9 789349 964235